Lecture Notes in Mathematics

Edited by A. Dold and B. Eckmann

1196

Eduardo Casas-Alvero
Sebastian Xambó-Descamps

The Enumerative Theory of Conics after Halphen

Springer-Verlag
Berlin Heidelberg New York London Paris Tokyo

Authors

Eduardo Casas-Alvero
Sebastian Xambó-Descamps
Department of Geometry and Topology, University of Barcelona
Gran Via 585, Barcelona 08007, Spain

Mathematics Subject Classification (1980): 14N10; 14C17; 51N15

ISBN 3-540-16495-2 Springer-Verlag Berlin Heidelberg New York
ISBN 0-387-16495-2 Springer-Verlag New York Berlin Heidelberg

Printing and binding: Beltz Offsetdruck, Hemsbach/Bergstr.
2146/3140-543210

Introduction.

This work deals on the one hand with understanding the contents of Halphen's contribution to the subject of enumerative theory of conics, and on the other with extending his theory to conditions of any codimension. The reader interested in the history of this subject may profit from the beautiful paper of Kleiman [K.2].

In the enumerative theory of conics there have been basically three approaches, namely those associated to De Jonquières, to Chasles, and to Halphen (see the works of these authors referred to in the references, as well as [K.2] and the references therein). Conceptually the first two are similar in that they correspond to computations performed in the Chow ring of $\mathbb{P}_5$ and of the variety of complete conics, respectively. Unfortunately the numbers obtained with these aproaches need not have enumerative significance, even if the data in the problem under consideration are in general position. A famous example of this failure is the answer given by De Jonquières theory to the problem of finding the number of conics that are tangent to five given conics in general position. Similarly, Halphen gave examples of this unsatisfactory situation, needless to say a little more involved, for the theory of Chasles (see [H.3.], §15, or the example 14.8 in this memoir).

On the other hand, the starting point in Halphen's theory is the distintion between proper and improper solutions (see §6) to an enumerative problem and his goal is to count the number of proper solutions. The numbers produced with this theory have always enumerative significance in the sense that if the data of the (reduced) conditions involved are in general position, then such numbers always are the number of distinct proper solutions of the problem. In addition, it turns out that all nondegenerate solutions are proper solutions and, if the data of the conditions are in general position, then, conversely, all proper solutions are non-degenerate, so that for (reduced) conditions with data in general position Halphen's theory gives the number of non-degenerate solutions.

In relation to this last point we should say that recently De Concini and Procesi [D-P] have taken the number of non-degenerate solutions, in the general setting of symmetric spaces, as the corner stone for an abstract enumerative theory.

The present work is the result of a project begun about two years ago by the first author with the idea of understanding Halphen's results and of providing modern proofs for them. This took about one year and the output was roughly the contents of §§ 1-14. Halphen considered two kinds of enumerative problems, namely, (1) to find the number of conics in a one-dimensional system that properly satisfy a given first order condition, and (2) to find the number of conics properly satisfying five independent conditions. These problems are solved by what we call Halphen's first and second formula, which are the contents of Theorem 9.2 and 14.6.

Although the basic ideas of this first part are due to Halphen, the presentation and many of the proofs are new. This is especially so for the definition of local characteristic numbers of first order conditions and the proof of Halphen's first formula.

After this first part had taken shape we became interested in finding analogues to these ideas for conditions of any codimension. The joint work in this direction has been developed in the last twelve months and the results are the contents of §§ 15-23. In spite of the fact that the results of the first part can be obtained again from results of the second, we have nevertheless maintained the two parts in order to offer, in the first, a rather elementary and updated version of Halphen's work on the subject, and, in the second, a general treatment for conditions of any codimension.

Finally we give a brief description of the contents of §§ 15-23. Section 15 is devoted to recall the structure of the Chow ring of the variey **W** *of complete conics and to list a number of cycles and relations among them which are needed later on.*

Sections 16, 17 and 18 are more general than the rest and are devoted to prove a generalization of the classical formula of Noether about the intersection of plane curves (see theorem 16.6) and to use it for a generalization of Halphen's first formula (see theorem 18.5).

Sections 19 and 20 are devoted to the construction of certain conditions (cycles), to the definition of strict equivalence of conditions and the groups $\mathrm{Hal}^{\bullet}(\mathbf{W})$, *and to prove that the strict equivalence classes of those cycles provide a free* $\mathbb{Z}$*-basis for* $\mathrm{Hal}^{\bullet}(\mathbf{W})$*. The main tool here is a particularization of Halphen's generalized formula (18.5) to the case of conics and a numerical criterion for strict equivalence proved with the resulting formula (20.2 and theorem 20.4).*

In section 21 the graded group Hal$^{\bullet}$(**W**) *is given a structure of graded commutative ring with unit, the enumerative significance of which is explained in theorem 21.7. This ring is the abstract ring of De Concini and Procesi in the case of conics. The product in* Hal$^{\bullet}$(**W**) *is made explicit in section 22 by showing how to compute the products of any two terms of the basis constructed before. In section 23 we work out two examples.*

E. Casas Alvero

S. Xambó Descamps

List of notations and conventions

$\mathbb{Z}$	ring of integers
$\mathbb{Q}$	field of rational numbers
$\mathbb{R}$	field of real numbers
$\mathbb{C}$	field of complex numbers
R^*	group of units of the ring R
$\mathbb{P}_n$	n-dimensional projective space over $\mathbb{C}$
$\check{\mathbb{P}}_n$	dual space of $\mathbb{P}_n$, or projective hyperplane space
$\mathrm{ord}_x(f)$	least exponent of the non-zero terms of the (broken) power series f
$\mathcal{O}_{X,Z}$	local ring of the irreducible subvariety Z of an algebraic variety X

If $Z \subseteq Y \subseteq X$, Y defines an ideal in $\mathcal{O}_{X,Z}$ that we shall call "ideal of Y in $\mathcal{O}_{X,Z}$".

Sing(X)	the singular set of an algebraic variety X
$D \sim D'$	divisors D and D' are linearly equivalent
$\sigma(K)$, K^σ	the result of transforming K by an element σ of a group acting on a set of objects where K belongs.

Given cycles K and K' on a smooth algebraic variety X, and a component Z of $|K| \cap |K'|$, where $|K|$ means the support of K, if $\dim(Z) = \dim(K) + \dim(K') - \dim(X)$, then Z will be called a proper component of the intersection $K \cap K'$ and the intersection multiplicity of K and K' at Z will be denoted by $i_Z(K \cdot K')$. The variety X will be clear from the context. If K and K' meet properly, i.e., all components of $K \cap K'$ are proper, then $K \cdot K'$ will denote their intersection cycle, i.e., $\sum i_Z(K \cdot K') \cdot Z$, the summation extended over all components of the intersection.

Given a zero cycle K on X, and a subset V of X, $\#_V K$ will mean the sum of the multiplicities in K of the points of V. Instead of $\#_X K$ we will also write $\int_X K$ or deg(K). If X is complete and α is a rational class of 0-cycles, $\int_X \alpha$ will denote the degree for any cycle representing α.

Suppose K and K' are such that $\dim(K)+\dim(K') = \dim(X)$ and that K and K' intersect properly. Then

$$\int_X (K \cdot K')$$

will also be denoted simply by $(K \cdot K')$, or $K \cdot K'$ if no confusion should arise, and will be called (total) intersection number of K and K'.

$A^{\bullet}(X)$ will denote the intersection ring of X, graded by codimension. The class of a cycle K in $A^{\bullet}(X)$ will be denoted by $[K]$.

If X and X' are smooth projective varieties and $f: X \longrightarrow X'$ is a morphism, $f_*: A^{\bullet}(X) \longrightarrow A^{\bullet}(X')$ and $f^*: A^{\bullet}(X') \longrightarrow A^{\bullet}(X)$ will denote the usual push-forward and pull-back maps associated to f.

CONTENTS

§ 1. Conics.

We shall reserve the term *conic* to mean a curve of degree 2 in $\mathbb{P}_2$ (the projective plane over the field $\mathbb{C}$ of complex numbers). As a reference for conics see [S-K]. Here we recall a few basic facts for convenience of the reader. With respect to a projective system of coordinates ξ_o, ξ_1, ξ_2 of $\mathbb{P}_2$, a conic is given by an equation of the form

$$(1) \qquad \sum_{i,j} a_{ij}\, \xi_i\, \xi_j = 0, \qquad i,j=0,1,2,$$

where $a_{ij} \in \mathbb{C}$, $a_{ij}=a_{ji}$, and not all a_{ij} vanish. The coefficients a_{ij} determine the same conic as the coefficients b_{ij} if and only if there exists $\lambda \in \mathbb{C}^*$ such that $b_{ij} = \lambda a_{ij}$. In other words, the matrix $A = (a_{ij})$ is symmetric and determined up to a non-zero scalar factor by the conic; we will say that A is the *conic matrix* . Therefore the set of conics is in a one-to-one correspondence with the projective space $\mathbb{P}_5$ associated to the vector space of symmetric 3×3 matrices. The projective structure induced on the set of conics via this correspondence is independent of the coordinate system.

A *linear system* of conics is a linear subspace of $\mathbb{P}_5$. A *pencil* of conics is a 1-dimensional linear system of conics.

A conic (1) is irreducible (or *non-degenerate*) if and only if $\det(a_{ij}) \neq 0$. An irreducible conic is smooth and rational (isomorphic to $\mathbb{P}_1$), whereas a degenerate conic (i.e., a conic such that $\det(a_{ij})=0$) is a pair of lines (if $\operatorname{rank}(a_{ij}) = 2$) or a double line (if $\operatorname{rank}(a_{ij}) = 1$). The equation $\det(a_{ij}) = 0$ defines a cubic hypersurface $D \subset \mathbb{P}_5$, the hypersurface of degenerate conics.

Let $\check{\mathbb{P}}_2$ denote the dual of $\mathbb{P}_2$, so that $\check{\mathbb{P}}_2$ is the projective plane of lines of $\mathbb{P}_2$ Then the image V of $\check{\mathbb{P}}_2$ in $\mathbb{P}_5$ by the *Veronese map* , which sends the line u: $u_o\xi_o + u_1\xi_1 + u_2\xi_2 = 0$ to the double line $u^2=0$, is the Veronese surface in $\mathbb{P}_5$. As usual we will take (u_o,u_1,u_2) as projective coordinates of u and will identify $\mathbb{P}_2$ to $\check{\check{\mathbb{P}}}_2$.

We will set $\bar{a}_{ij}$ to denote the cofactor of a_{ij} in the matrix (a_{ij}). Then $\bar{a}_{ij}=0$, $0 \leq i,j \leq 2$, are equations for V with its reduced structure (see [S], Ch. I, §4). Moreover, V is equal to the singular set of D, whereas D is the chord variety of V, as an elementary computation shows.

By a *conic envelope* we will understand a conic in $\check{\mathbb{P}}_2$. The 5-dimensional projective space of conic envelopes can be identified with $\check{\mathbb{P}}_5$, the dual of the $\mathbb{P}_5$ of conics, through the apolarity relation

$$\sum_{i,j} a_{ij}\, \alpha_{ij} = 0 \tag{2}$$

between a conic and a conic envelope given by matrices (a_{ij}) and (α_{ij}), respectively. Thus each conic envelope can be identified with the 4-dimensional linear system of those conics which are apolar to it. If for instance the conic envelope is a pair of points P, Q, then the linear system of apolar conics to this pair is the set of conics that harmonically divide PQ and if the conic envelope is a double point P, then it is the system of conics that go through P.

By $\check{D}$ we will denote the cubic hypersurface of $\check{\mathbb{P}}_5$ whose points are degenerate conic envelopes, and by $\check{V}$ the Veronese surface of double points, so that $\check{V}$ is the singular set of $\check{D}$ and $\check{D}$ is the chord variety of $\check{V}$.

The dual of a non-degenerate conic is a non-degenerate envelope, and conversely. This gives an isomorphism $\mathbb{P}_5-D \xrightarrow{\sim} \check{\mathbb{P}}_5-\check{D}$. If A is the matrix of a non-degenerate conic, then the associated conic envelope has matrix A^{-1}, or, equivalently, the matrix $\bar{A}$ of cofactors of A.

§ 2. Complete conics

The traditional point of view in projective geometry has been to consider the plane of points $\mathbb{P}_2$ and the plane of lines $\check{\mathbb{P}}_2$ simultaneously, and to think each as the dual of the other. It is in this sense that the classical geometers, when thinking of a (non-degenerate) conic, really understood it as a pair formed by the conic and its line envelope and regarded each of these as an aspect of the conic.

In the non-degenerate case, the consideration of its conic evelope adds no information to the given conic. It is only when dealing with degenerate conics (usually considered as limits of non-degenerate ones) that the simultaneous consideration of a point conic and one of its envelopes contains more information than the conic alone, as this does not determine uniquely the envelope nor conversely. Conics, when considered in this double aspect as locus of points and evelope of lines, are called, since Van der Waerden's work [W], *complete conics*.

To give a precise definition, consider the duality isomorphism $\alpha : \mathbb{P}_5-D \xrightarrow{\sim} \check{\mathbb{P}}_5-\check{D}$ which transforms a conic locus into its conic envelope. Set $\mathbf{W}$ to denote the clousure in $\mathbb{P}_5 \times \check{\mathbb{P}}_5$ of the graph $\mathbf{W}_o$ of α. Then $\mathbf{W}$ is called the variety of complete conics. The elements of $\mathbf{W}_o$ are called non-degenerate complete conics; the elements of $\mathbf{W}-\mathbf{W}_o$ are referred to as degenerate complete conics. Let $p: \mathbf{W} \longrightarrow \mathbb{P}_5$ be the restriction of the first projection, and $t: \mathbf{W} \longrightarrow \check{\mathbb{P}}_5$ the restriction of the second projection. Then $p^{-1}(\mathbb{P}_5-D) = t^{-1}(\check{\mathbb{P}}_5-\check{D}) = \mathbf{W}_o$ and $p: \mathbf{W}_o \xrightarrow{\sim} \mathbb{P}_5-D$, $t: \mathbf{W}_o \xrightarrow{\sim} \mathbb{P}_5-\check{D}$, so that in particular p and t are birrational isomorphisms. For a given $C \in \mathbf{W}$, we will say that $p(C)$ is the *conic locus* of C and that $t(C)$ is the *conic envelope* of C.

Since $\alpha : \mathbb{P}_5-D \longrightarrow \mathbb{P}_5-\check{D}$ can be viewed as the map which transforms a non-degenerate 3×3 symmetric matrix $a=(a_{ij})$ into the matrix $\bar{a} = (\bar{a}_{ij})$

of cofactors of a, we see that α is actually regular on the open set $\mathbb{P}_5 - V$. In fact $p: \mathbf{W} \longrightarrow \mathbb{P}_5$ can be identified with the blowing up of $\mathbb{P}_5$ along V. Dually, $t: \mathbf{W} \longrightarrow \check{\mathbb{P}}_5$ can be identified with the blowing up of $\check{V}$ (the Veronese surface of double points). In particular $\mathbf{W}$ is a smooth irreducible projective variety.

Let us denote a_{ij} the coordinates of $\mathbb{P}_5$, as before, and α_{ij} the dual coordinates of $\check{\mathbb{P}}_5$. Then the points of $\mathbf{W}$ satisfy the equations got eliminating ρ in the relation

$$(a_{ij})(\alpha_{ij}) = \rho I, \tag{1}$$

where I is the 3×3 identity matrix, or, in other words, the relations obtained setting the non diagonal entries of $(a_{ij})(\alpha_{ij})$ equal to zero and equating the three diagonal entries of the same matrix. In particular, for the degenerate complete conics $\rho = 0$, so that they satisfy the relation $(a_{ij})(\alpha_{ij}) = 0$. These relations imply immediately the following statements:

2.1. If the conic locus of a complete conic is a pair of (distinct) lines, then its conic envelope is the common point of the two lines, counted twice. Dually, if the conic envelope of a complete conic is a pair of (distinct) points, then its conic locus is the line joining the two points, counted twice.

2.2. If the conic locus of a complete conic is a double line, then its conic envelope is a pair of points (not necessarily distinct) on the line. Dually, if the conic envelope of a complete conic is a double point, then its conic locus is a pair of lines (not necessarily distinct) through the point.

Conversely we have the following statements:

2.3. A pair consisting of a double line and a pair of points (possibly equal) on it is a complete (degenerate) conic.

Dually, a pair consisting of a pair of lines (possibly equal) and a common point, counted twice, is a complete conic.

Proof

Let L be a line and $P,Q \in L$. First assume that $P \neq Q$. In this case consider a pencil of conics bitangent at P and Q. Then $\{L,\{P,Q\}\}$ belongs to the closure of the 1-dimensional rational family of non-degenerate conics in the pencil. If $P=Q$ one uses a four point contact pencil with base point P and fixed tangent L. □

Actually it turns out that the relations before are equations for the subvariety $\mathbf{W}$ of $\mathbb{P}_5 \times \check{\mathbb{P}}_5$.

Now that we have described the degenerate complete conics we will fix some notation. Consider the map $p\colon \mathbf{W} \longrightarrow \mathbb{P}_5$, which we know to be the blowing up of $\mathbb{P}_5$ along the Veronese surface V of double lines. Thus we see that the points of $p^{-1}(V)$, the exceptional variety, are the complete conics consisting of a double line and a pair of points on it. We will set $\check{\mathbf{A}} = p^{-1}(V)$. In a similar way, $\mathbf{A} := t^{-1}(V)$ is the exceptional variety of the blowing up $t\colon \mathbf{W} \longrightarrow \check{\mathbb{P}}_5$ of $\check{\mathbb{P}}_5$ along $\check{V}$. By the general properties of the blowing up, $\mathbf{A}$ and $\check{\mathbf{A}}$ are irreducible and smooth. Clearly, $p(\mathbf{A}) = D$ and $t(\check{\mathbf{A}}) = \check{D}$, so that $\mathbf{A}$ and $\check{\mathbf{A}}$ are the strict transforms of D and $\check{D}$ under the blowing up maps p and t, respectively.

In the sequel we will set $\mathbf{B}$ to denote the intersection of $\mathbf{A}$ and $\check{\mathbf{A}}$. As we will see, $\mathbf{B}$ is irreducible and smooth. Moreover, $\mathbf{A}$ and $\check{\mathbf{A}}$ meet transversally along $\mathbf{B}$. This allows to distinguish the following three types of complete degenerate conics:

Type A: The conics in A–B, i.e., two distinct lines with their common point counted twice.

Type Ǎ: The conics in Ǎ–B, i.e., a double line with two distinct points on it, usually called foci of the degenerate conic.

Type B: The conics in B, i.e., a double line with a double point on it, called the double focus of the complete conic.

Types A and Ǎ are dual of each other, while type B is seld-dual.

2.4. Definitions

Let $C=(c,\check{c})$ be a complete conic. Then we will say that C goes through a given point P iff $P \in c$; that C is tangent to a line u, iff $u \in \check{c}$; that C cuts on the line u the two points P, Q iff $u \cap c = \{P,Q\}$; that C has tangents u, v from a given point P, iff $P^* \cap \check{c} = \{u,v\}$, where P^* is the pencil of lines going through P; that C is apolar with a conic envelope $\bar{c}^\vee$, iff c and $\bar{c}^\vee$ are apolar; and that C is apolar with a conic locus $\bar{c}$, iff $\bar{c}$ and $\check{c}$ are apolar.

Now let (a_{ij}) be the matrix of c and (α_{ij}) the matrix of $\check{c}$. Given a point X and a line u we will set (x_o,x_1,x_2) and (u_o,u_1,u_2) to denote their coordinates. Then two points X, X' are said to be *conjugate* with respect to C iff $\sum a_{ij}x_ix'_j = 0$, or equivalently, iff the pair of points that c cuts on the line $X X'$ harmonically separates the pair X, X' (in case $X=X'$ this means that c goes through X). The points which are conjugate of any other point are said to be *double points* of C; they are the points satisfying the relation $(x_o,x_1,x_2)(a_{ij}) = (0,0,0)$. The set of double points is empty if C is non-degenerate, is reduced to the common point of the two lines in case of conics of type A, and coincides with the set of points on the (double) line for conics of type $\check{A}$ or B.

Dually, two lines u,u' are said to be *conjugate* with respect to C iff $\sum \alpha_{ij}u_i u'_j = 0$. This is equivalent to assert that the pair of tangents to C from the common point of u and u' harmonically separates the pair u,u' (if $u=u'$, this simply means that u is tangent to C). A line is said to be a double tangent if it is conjugate to any other line; they satisfy the relation $(u_o,u_1,u_2)(\alpha_{ij}) = 0$. Thus the set of double tangents is empty if the conic is non-degenerate, is reduced to the (double) line in conics of type $\check{A}$, and is the pencil of lines through the double point in conics of types A and B.

Let X be a point which is not a double point of C. Then the points X' which are conjugate to X lie on a line u given by the relations $(u_o,u_1,u_2) = (x_o,x_1,x_2)(a_{ij})$. This line is called the *polar line* of X with respect to C. Dually, if u is a line which is not a double tangent, then the lines u' which are conjugate to u with respect to C pass through a fix point X given by the relations $(x_o,x_1,x_2) = (u_o,u_1,u_2)(\alpha_{ij})$. This point is called the *pole* of u with respect to C. For double points (resp. lines) the notion of polar line (resp. pole) is not defined.

Let X be a point which is not a double point of C, and let u be a line which is not a double tangent of C. Let u' and X' be the polar line of X and the pole of u, respectively. Then from the relation (1) in this § we get that $\rho(x_o,x_1,x_2) = (u_o,u_1,u_2)(\alpha_{ij})$ and $\rho(u_o,u_1,u_2) = (x_o,x_1,x_2)(a_{ij})$. Therefore we see that if C is non-degenerate (i.e., $\rho \neq 0$), then the pole of u' is X and the polar line of X' is u, so that the relation between pole and polar is one-to-one and symmetrical. But if C is degenerate, then u' is a double tangent and X' a double point for all X and u as above.

A triangle is said to be self-polar with respect to C if the polar of each vertex which is not a double point of C is equal to the opposite side, and the pole of each line which is not a double tangent is equal to the opposite vertex.

The set of conic loci for which a given triangle τ is self-polar are the points of a plane S_τ in $\mathbb{P}_5$. If the triangle is taken as a system of coordinates, then the equations of the plane are simply $a_{ij} = 0$, $i \neq j$. It is easy to see that the complete conics which admit τ as a self-polar triangle are the points of a smooth surface $\tilde{S}_\tau$ in $\mathbf{W}$ which coincides with the blowing up of S_τ at the three points corresponding to the three sides of the triangle, each counted twice.

The last observation can be used to show:

2.5. Lemma

The hypersurfaces $\mathbf{A}$ and $\check{\mathbf{A}}$ meet transversally along $\mathbf{B}$.

Proof

Let C be an element of $\mathbf{B}$, so that C consists of a line u and a point P, both counted twice. Choose any triangle τ in $\mathbb{P}_2$ such that u is a side of τ and P a vertex. Let $\tilde{S}_\tau \subset \mathbf{W}$ be the surface of complete conics for which τ is self-polar. Then it is enough to see that $\mathbf{A} \cap \tilde{S}_\tau$ and $\check{\mathbf{A}} \cap \tilde{S}_\tau$ meet transversally at C. But this is clear if one uses the fact that $p\colon \tilde{S}_\tau \longrightarrow S_\tau$ is the blowing up of S_τ at three points.

Remark

In the sequel, unless otherwise stated, the elements of $\mathbf{W}$ will be referred to as conics, the elements of $\mathbb{P}_5$ as conic loci, and the elements of $\check{\mathbb{P}}_5$ as conic envelopes.

§3. Systems of conics

By a (1-*dimensional*) *system of conics* we will understand a reduced curve Γ in W such that no component of Γ is contained in **A** or $\check{\mathbf{A}}$. The system is said to be irreducible (resp. rational) if Γ is irreducible (resp. rational).

Let Γ be a system of conics. Then $p(\Gamma)$ and $t(\Gamma)$ are curves in $\mathbb{P}_5$ and $\check{\mathbb{P}}_5$, respectively. We will say that $p(\Gamma)$ is the system of conic loci, and $t(\Gamma)$ of conic envelopes, associated to Γ. Notice that since neither **A** nor $\check{\mathbf{A}}$ contain components of Γ, both $p(\Gamma)$ and $t(\Gamma)$ have the same number of components as Γ. Furthermore, if Γ is irreducible then $p: \Gamma \longrightarrow p(\Gamma)$ and $t: \Gamma \longrightarrow t(\Gamma)$ are birrational morphisms.

3.1. Definition

The integers $\mu = \deg p(\Gamma)$, $\nu = \deg t(\Gamma)$ will be called *characteristic numbers of* Γ.

3.2. Proposition

Let Γ be a system of conics and let (μ,ν) be its pair of characteristic numbers. Then μ is the number of conics in Γ that pass through a generic point of $\mathbb{P}_2$. Dually, ν is the number of conics in Γ which are tangent to a generic line of $\mathbb{P}_2$.

Proof

The conic loci that go through a point P form a hyperplane H_P in $\mathbb{P}_5$, so that if $H_P \not\supset \Gamma$ then μ is the number of conics in Γ going through P, each counted with a suitable multiplicity. Therefore it will be enough to show that if P is generic in $\mathbb{P}_2$ then H_P is not tangent to $p(\Gamma)$. Now the set of hyperplanes H_P is the Veronese surface $\check{V} \subset \check{\mathbb{P}}_5$, so that if follows that if all these H_P are tangent to $p(\Gamma)$ then there exists a component Γ' of $p(\Gamma)$ such that for each point c of Γ' there are ∞^1 hyperplanes H_P tangent to Γ' at c. Thus, if c is generic on Γ', the tangent line T to Γ' at c, which is a pencil of conics, is contained in ∞^1 hyperplanes of the form H_P. This means that the pencil T has infinitely many base points, which is only possible if all conics in T, in particular c, are degenerate. But this implies that $p(\Gamma)$ has a component whose points are all degenerate conics. Since this contradicts the definition of system of conics, the proof is complete. □

3.3. Remark

It is clear that μ also coincides with the number of conics in Γ that harmonically divide a pair of generic points, or that are apolar with a generic envelope. Dually, ν is the number of conics in Γ whose tangents through a generic point harmonically divide a pair of generic lines through that point. In the case $\mu=1$ (resp. $\nu=1$) the system is called a *pencil of conics* (resp. a *range of conics*). Both are irreducible and rational.

By definition, in a system Γ of conics there are at most a finite number of degenerate conics. According to the classical view which regards degenerate conics as limits of non-degenerate ones[(1)], we define a *degeneration* of Γ as a branch of Γ centered at a degenerate conic. We will say that a degeneration is of type A, $\check{A}$ or B according to whether its center is respectively of type A, $\check{A}$ or B.

3.4. Definition

Let γ be a degeneration of a system Γ. Then the *local characteristic numbers* of γ, or of Γ at γ, are the pair of non-negative integers (m,n) defined as the intersection multiplicities of γ with $\check{\mathbf{A}}$ and $\mathbf{A}$, respectively. We will also say that m is the *order* of γ and n the *class* of γ. A branch of Γ whose center is a non-degenerate conic will be assigned order and class both equal to 0. It is clear that a degeneration is of type A iff $m=0$ and $n>0$; of type $\check{A}$ iff $m>0$ and $n=0$; and of type B iff

(1) "Il peut se trouver des figures qui ne soient pas des coniques, mais des limites de coniques" ([H1], p.6).

$m > 0$ and $n > 0$.

3.5. Remark

$M := (\Gamma.\check{A})$ and $N := (\Gamma.A)$ are, respectively, the sum of the orders and classes of all degenerations of Γ.

§4. Equations of the degeneration hypersurfaces

Let $(P_o, P_1, P_2; Q)$ be a projective system of coordinates and let U_2 denote the open set in **W** of conics which are not tangent to $u_2 := P_oP_1$. It is clear that a conic is in U_2 iff $\alpha_{22} \neq 0$, so that we will also write $U_2 = D(\alpha_{22})$. Now any complete conic satisfies the relations (1) of §2 which imply the relations $\alpha_{22}\bar{a}_{ij} = \alpha_{ij}\bar{a}_{22}$, so that for conics in U_2

$$\bar{a}_{ij} = \frac{\alpha_{ij}}{\alpha_{22}}\,\bar{a}_{22} \quad .$$

Since $\bar{a}_{ij}=0$ are equations for $\check{A}$, it turns out that for any form F of degree 2 in the a_{ij} the rational function $\bar{a}_{22}/F$ is a local equation for $\check{A}$ in the open set $U_2 \cap D(F)$. On the other hand $\bar{a}_{22}$ vanishes on $\check{L}_{u_2} := \mathbf{W}-U_2$, the hypersurface of conics which are tangent to u_2. It follows that the divisor of zeroes of $\bar{a}_{22}$ has $\check{A}$ and $\check{L}_{u_2}$ as its components, both counted once. $\check{L}_{u_2}$ counts once because in a neighbourhood of a non-degenerate conic $\bar{a}_{22}$ can be used, instead of α_{22}, as a local equation of $\check{L}_{u_2}$. The hypersurfaces $\check{A}$ and $\mathbf{W}-U_2$ have as intersection the variety of double lines with one of its foci on u_2.

Suppose now that u is a line and that P_o, P_1, Q_2 are three distinct points on u. Let C be a conic not going through P_o and let Z, Z' be the points at which C meets u. Define the function

$$Y = Y(C) := [(P_o,P_1,Q_2,Z) - (P_o,P_1,Q_2,Z')]^2,$$

where (a,b,c,d) means the cross ratio of a,b,c,d. Then one has:

4.1. Proposition

Y is rational function on $\mathbf{W}$ which is regular on the open set of conics not going through P_o. Moreover, it is a local equation of $\check{A}$ in a neighbourhood of any conic not through P_o and not tangent to u.

Proof

We may choose a projective system of coordinates whose first points are P_o and P_1 and such that Q_2 is the projection of the unit point Q on u from the third point P_2. Then the coordinates (z_o,z_1,z_2) of the points in which C meets u are those satisfying the relations

$$x_2 = 0, \qquad a_{oo}x_o^2 + 2a_{o1}x_ox_1 + a_{11}x_1^2 = 0.$$

To say that C does not go through P_o is equivalent to say that $a_{oo}\neq 0$. This implies that $x_1\neq 0$ and so $t = x_o/x_1$ satisfies the quadratic equation $a_{oo}t^2 + 2a_{o1}t + a_{11} = 0$. If t_1,t_2 are the roots of this equation, then it is clear that

$$Y(C) = (t_1-t_2)^2 = (t_1+t_2)^2-4t_1t_2 = \left(\frac{-2a_{o1}}{a_{oo}}\right)^2 - 4\,\frac{a_{11}}{a_{oo}} = 4\,\bar{a}_{22}/a_{oo}^2 ,$$

which is a rational function on $\mathbf{W}$. The second statement of the proposition is a direct consequence of this expression of Y and the considerations at the beginning of this §. □

Dually, let P be a point in $\mathbb{P}_2$ and pick three distinct lines u_o,u_1,u through P. Given a conic C not tangent to u_o, let v,v' be the tangents to C drawn from P. Define the function

$$X = X(C) := [(u_o,u_1,u,v) - (u_o,u_1,u,v')]^2$$

Then one has

4.1ˇ. Proposition

X is a rational function on **W** which is regular on the open set of conics which are not tangent to u_o. Moreover, X is a local equation of **A** in a neighbourhood of any conic which is not tangent to u_o and does not go through P. □

4.2. Remarks

(a) From the expression of Y obtained in the proof of proposition 4.1 it follows that its divisor of poles on **W** is twice the hypersurface of conics through P_o, whereas its divisor of zeroes has two components, each counted once — **Ǎ** and the hypersurface of conics tangent to u. The points of indeterminacy of Y are double lines which pass through P_o and conics which are tangent to u at P_o. Dually, the polar divisor of X is twice the hypersurface of conics tangent to u_o, and its divisor of zeroes has two components, each counted once — **A** and the hypersurface of conics going through P. The points where X is indeterminate are the conics which are tangent to u_o at P and the pairs of lines with its double point at P.

(b) From propostions 4.1 and 4.1ˇ it turns out that X,Y are local equations of **B** in a neighbourhood of any conic which does not go through P or P_o and is not tangent to u or u_o. In particular, and for the sake of simplicity, in the sequel we take $P=P_o$ and $u=u_o$, in which case X,Y are local equations of **A** and **Ǎ**, respectively, in the neighbourhood of any conic which does not go through P and is not tangent to u.

(c) Given P and u a change in the points of u used to define Y, or in the lines through P used to define X, only changes Y, or X, into λY, or μX, where λ(or μ) is a non-zero scalar. □

We end this section stating a direct corollary of propositions 4.1 and $4.1^{\vee}$.

4.3. Corollary

Let $C = C(s)$ be a parametrization of a degeneration γ of a system of conics Γ. Let (m,n) be the pair of characteristic numbers of γ. Assume that $C(0)$ is not tangent to the line u and does not go through the point P used to define X and Y. Then

$$m = \mathrm{ord}_s\, Y(C(s)), \qquad n = \mathrm{ord}_s X(C(s)). \quad \square$$

§5. Conditions imposed on conics

An effective codimension i cycle of **W** shall be called a *condition of order* i (imposed on conics). In this section we will deal with conditions of order 1, i.e., effective divisors of **W**, which for simplicity will be called *conditions*. The first step will be to define local and global characteristic numbers for such conditions.

Let us start recalling the structure of the group $\mathrm{Pic}(\mathbf{W})$. Since $p\colon \mathbf{W} \longrightarrow \mathbb{P}_5$ is the blowing up of $\mathbb{P}_5$ along the Veronese surface V and $\check{A} = p^{-1}(V)$, $p^*\colon \mathrm{Pic}(\mathbb{P}_5) \longrightarrow \mathrm{Pic}(\mathbf{W})$ is a monomorphism, $\mathbb{Z}\cdot[\check{A}]$ is infinite cyclic, and

$$\mathrm{Pic}(\mathbf{W}) = p^*\mathrm{Pic}(\mathbb{P}_5) \oplus Z\cdot[\check{A}] \quad ,$$

where $[X]$ denotes the class in $\mathrm{Pic}(\mathbf{W})$ of a divisor X ([Har], Ch. II, Ex. 8.5). Now Pic $(\mathbb{P}_5)$ is infinite cyclic generated by a hyperplane, hence $p^*\mathrm{Pic}(\mathbb{P}_5)$ is infinite cyclic generated by the class L of the inverse image of a hyperplane. In the sequel, unless some confusion could arise, instead of denoting [X] the class of a divisor X we will write X for short. Sometimes it will be enough to comment whether a statement is to be considered as a cycle relation or a relation between cycle classes. In any event, $X \sim Y$ will denote that X and Y are linearly equivalent, or, if Y is a class already, that X is in the class Y.

The same argument applied to $t: \mathbf{W} \longrightarrow \check{\mathbb{P}}_5$ shows that $\mathrm{Pic}(\mathbf{W})$ is as well a free abelian group with generators $\check{L}$ and $\mathbf{A}$, where $\check{L}$ is the class of the inverse image under t of a hyperplane of $\check{\mathbb{P}}_5$.

5.1. Proposition

In Pic ($\mathbf{W}$) the following relations hold:

$$\mathbf{A} \sim 2\check{L}-L, \qquad \check{\mathbf{A}} \sim 2L-\check{L} .$$

From this relations it follows that $\mathrm{Pic}(\mathbf{W})$ is free abelian on the generators L and $\check{L}$.

Proof

Since the hypersurface $D \subset \mathbb{P}_5$ of degenerate conic loci has degree 3, $p^*(D) \sim 3L$. On the other hand, since the strict transform of D under p is $\mathbf{A}$ and since $\mathrm{Sing}(D) = V$, it turns out that $p^*D = \mathbf{A} + 2\check{\mathbf{A}}$ (this may also be directly verified). Thus $\mathbf{A} + 2\check{\mathbf{A}} \sim 3L$. Dually, $2A + \check{A} \sim 3\check{L}$. From these the stated relations are easily deduced. □

5.2. Definition

Given a condition K, if $K \sim \alpha L+\beta\check{L}$, $\alpha,\beta \in \mathbb{Z}$, then the integers α,β will be called *global characteristic numbers* of K.

5.3. Proposition

Let K be an irreducible hypersurface of $\mathbf{W}$ and assume $K \sim \alpha L+\beta \check{L}$. Then if $K \neq \check{\mathbf{A}}$ (resp. $K \neq \mathbf{A}$), $\deg p_*(K) = \alpha+2\beta$ (resp. $\deg t_*(K) = 2\alpha+\beta$).

Proof

In fact $p_*(L)$ is the class of a hyperplane in $\mathbb{P}_5$ and $p_*(\check{L}) =$ $= p_*(2L-\check{A}) = 2p_*(L)$. □

5.4. Examples of conditions

(a) Conditions $K \sim L$ have global characteristic numbers (1,0). They will be called *point-linear* conditions. Among them we have the condition L_P of going through a point P, the condition of harmonically dividing a pair of points, or the condition of being apolar with a fixed envelope.

($\check{a}$) Conditions $K \sim \check{L}$ have global characteristic numbers (0,1). They will be called *tangentially linear*. Among them we have the condition $\check{L}_u$ of being tangent to a line u.

(b) A is the condition "to have a double point as envelope component"; its global characteristic numbers are (-1,2).

($\check{b}$) Dually, $\check{A}$ is the condition "to have a double line as locus component"; its global characteristic numbers are (2,-1).

5.5. Theorem (Chasles' formula)

Let Γ be a system of conics and K a condition satisfied by finitely many conics of Γ. Let (μ,ν) and (α,β) be the global characteristic pairs of Γ and K, respectively. Then if we count each conic of Γ satisfying K with its intersection multiplicity, the number $\bar{N}$ of such conics is given by the formula

$$\bar{N} = \alpha\mu + \beta\nu \ .$$

Proof

By definition $\bar{N} = [\Gamma]\cdot[K]$ and $[K] = \alpha L+\beta\check{L}$, so that $\bar{N} = \alpha([\Gamma]\cdot L) +$

$\beta([\Gamma]\cdot\check{L})$. But $[\Gamma]\cdot L = \deg p(\Gamma) = \mu$ and $[\Gamma]\cdot L = \deg t(\Gamma) = \nu$. □

5.6. Corollary (Halphen)

Let M (resp. N) be the sum of the orders (resp. classes) of the degenerations of Γ. Then

$$M = 2\mu-\nu \quad \text{and} \quad N = 2\nu-\mu \quad .$$

Hence

$$3\mu = 2M+N \quad \text{and} \quad 3\nu = 2N+M \quad .$$

Proof

It is enough to observe that $M = (\Gamma\cdot\check{A})$, $N = (\Gamma\cdot \mathbf{A})$ and apply Chasles' formula. □

5.7. Examples

(a) Let Γ be the system of conics passing through four points in general linear position. Then $\mu=1$ because Γ is a pencil of conics. It is also clear that $M=0$. Therefore $\nu=2$ and $N=3$, so that there are two conics in Γ that are tangent to a line and three degenerate conics consisting each of a pair of lines.

(ǎ) Dually, for the range of conics Γ tangent to four lines in general linear position one obtains that $\mu=2$, $\nu=1$, $M=3$, and $N=0$.

(b) Let Γ be a pencil of bitangent conics. Then Γ has global characteristic numbers (1,1). Therefore $M=N=1$, so that Γ has a single degeneration of type A and a single degeneration of type Ǎ.

(c) Consider the system of conics passing through three given non-colinear points and which are tangent to a given general line. Then $\mu=2$ by example (a). Moreover, since there are no degenerations of type Ǎ, $M=0$ so that

$\nu=4$ and $N=6$.

(č) Dually, the system of conics tangent to three non-concurrent lines and going through a general point has the following characteristic numbers: $\mu=4$, $\nu=2$, $M=6$ and $N=0$.

§6. Action of the projective group and proper solutions of an enumarative problem

Let G be the group of linear projective transformations of $\mathbb{P}_2$, $G = PGL(\mathbb{P}_2)$. The group G acts on $\check{\mathbb{P}}_2$ and $G \xrightarrow{\sim} PGL(\check{\mathbb{P}}_2)$ under this action. The elements of G will be called *homographies*.

Since homographies transform conics in conics, G acts likewise on $\mathbb{P}_5$, $\check{\mathbb{P}}_5$, and $\mathbb{P}_5 \times \check{\mathbb{P}}_5$. Under this last action G leaves invariant $\mathbf{W}$, and so G also acts on the variety of complete conics. As is well known, the orbits of G under its action on $\mathbf{W}$ are as follows:

$\mathbf{W}_o = \mathbf{W}-(A\cup\check{A})$, the open set of non-degenerate conics;

$A-B$, the locally closed set of degenerate conics of type A;

$\check{A}-B$, the locally closed set of degenerate conics of type $\check{A}$; and

B, the closed set of degenerate conics of type B.

The action of G on $\mathbf{W}$ is essential for the distinction, according to Halphen's point of view, of proper solutions of an enumerative problem from improper ones. Actually, such a distinction and the consideration of only proper solutions in the computations are distinguishing features of Halphen's theory as compared with those preceeding it. Later on we will analyze more closely the differences between the theories of De Jonquières, Chasles, and Halphen, as well as the discordancies between their formulae.

6.1. Definition (Halphen)

Let K be a condition and C a conic. We will say that C *properly satisfies* K iff C satisfies K and there exists $\sigma \in G$ such that C does not satisfy $\sigma(K)$. Equivalently, C properly satisfies K iff $C \in |K|$ and not all points in the orbit of C under G satisfy K.

This definition calls for some comments. First of all notice that enumerative problems of conics are formulated in the frame of projective geometry (possibly using metric relations) and that they naturally undergo the action of G. This said, it is worthwhile to reflect on what the nature of conditions imposed to conics is. So far we have accepted to call (simple) conditions the hypersurfaces (or, more generally, the divisors) of W, without delving into the question of whether or not a relation of the form $C \in D$ really expresses a fact of a projective nature that may occur or not to C. From a purely projective point of view, to impose a condition to a conic means to force the conic to satisfy certain projectively invariant relation between the conic and some given configuration (the *datum* of the condition) in $\mathbb{P}_2$. With respect to a projective system of coordinates, the verification of this relation will be translated by the vanishing of one or more (simultaneous) invariants of the conic and the datum.

Now the point is that any condition in the former sense (i.e., a hypersurface of $\mathbf{W}$) may be considered in this way. In fact a system of bihomogeneous equations $F_r(a_{ij}, \alpha_{i'j'}) = 0$ can be understood as the expression of an invariant relation between a conic and the configuration of the elements of the system of projective coordinates. This relation may be intrinsically phrased in terms of cross ratios between elements of the conic and of the projective system of coordinates.

This understanding of conditions leads to the following view of the action of G on them. Let K be the condition which a conic C satisfies iff certain projectively invariant relation $R(C,F)$ between C and a datum F is verified. The projective invariance of R means that $R(C,F)$ is true iff $R(\sigma(C),\sigma(F))$ is true for any $\sigma \in G$, and so $\sigma(K)$ is the condition obtained impossing that a conic is in the relation R to $\sigma(F)$, the transformed datum.

We summarize this digression in three remarks.

6.2. Remarks

(a) The belonging of a conic C to a hypersurface K of $\mathbf{W}$ is a relation which is equivalent to a projectively invariant relation $R(C,F)$ between the conic and a certain configuration F in the plane (which may be the coordinate system).

(b) If K is interpreted in this way, then $\sigma(K)$ is the condition given by demanding that a conic C' be in the relation R to the configuration $\sigma(F)$.

(c) The conics which improperly satisfy a condition K described by $R(C,F)$ are the conics for which $R(C,\sigma(F))$ holds for all $\sigma \in G$, i.e., for "any position of the datum".

At this point one may ask for the existence of absolute conditions, i.e., conditions whose datum is empty. Clearly, these must be invariant hypersurfaces under G, so that, by the description of the orbits of G on W, we have:

6.3. Proposition

The only irreducible absolute (first order) conditions are $\mathbf{A}$ and $\check{\mathbf{A}}$. □

6.4. Definition

A condition K will be said to be *degeneration free* iff neither **A** nor $\check{\mathbf{A}}$ are components of K.

This definition is relevant insofar as we are not interested here in enumerating how many degenerate conics a system has. For these conditions the improper solutions have a very simple characterization:

6.5. Proposition (Halphen)

A conic C improperly satisfies a degeneration free condition K iff $C \in B$ and $K \supset B$.

Proof

If K is degeneration free, the only orbit under G that may be contained in K is B, so that from the last part of 6.1 the proof follows. □

Henceforth, and unless otherwise stated, we will deal only with degeneration free conditions.

We end this section with a description of an example of Halphen which this author used [H.1] to reject Chasles' formulation. We will maintain his using of metric (euclidian) terminology, more expressive than the usual projective phrasing that instead could be given.

6.6. Example

Let P be a point and u a line in the plane. Given a conic C, let $Y = Y(C)$ be the square of the length of the segment that C cuts on u, and $X = X(C)$ the square of the tangent of the angle between the tangents to C drawn from P. Consider the condition K defined by the relation $Y = X$. Then $B \subset K$ and so Chasles' formula does not give the number of proper solutions of K in a given system Γ if the latter has degenerations of type B.

§7. Genericity of the data of conditions

In enumerative geometry problems one cannot hope for precise results, and still having reasonably simple solutions, unless one disregards a multitude of particular cases whose description is too cumbersome. This is usually done by assuming an hypothesis of generality for the data of the conditions. Often this generality of the data is handled by means of the following theorem.

7.1. Theorem (Kleiman, [K.1])

Let U be an irreducible quasi-projective algebraic variety over $\mathbb{C}$ and assume that an algebraic group G acts transitively on U. Let V_1, V_2 be two equidimensional subvarieties of U. Then

(a) There exists a non-empty open set G' of G such that $\sigma(V_1) \cap V_2$ is either empty or has pure dimension $\dim V_1 + \dim V_2 - \dim U$ for all $\sigma \in G'$, and

(b) If V_1 and V_2 are smooth, then G' can be choosen in such a way that $\sigma(V_1) \cap V_2$ is smooth.

7.2. Corollary (Halphen [H.1], § 24)

Let Γ be a system of conics and K a condition. For σ generic in G, the conics in Γ which properly satisfy $\sigma(K)$ are non-degenerate, finite in number, and if K is reduced each appears with multiplicity 1 in the intersection $\sigma(K) \cap \Gamma$.

Proof

Since the first two assertions are satisfied for K iff they are satisfied for K_{red}, we may assume that K itself is reduced. Let U be one of the orbits $A-B$, $\check{A}-B$, or B and set $V_1 = K \cap U$, $V_2 = \Gamma \cap U$. If $V_1 \neq U$, then by the theorem $\sigma(V_1) \cap V_2 = \emptyset$ for $\sigma \in G$ generic, which implies that $\sigma(K) \cap \Gamma$ does not contain conics in the orbit U. If $V_1 = U$ then $K \supseteq U$ and this only can happen if $U=B$ (since conditions are assumed to be degeneration free), in which case the conics of B improperly satisfy $\sigma(K)$ and so Γ does not contain conics of B which properly satisfy $\sigma(K)$. Therefore for $\sigma \in G$ generic the proper solutions of $\sigma(K) \cap \Gamma$ are in W_o, the orbit of non-degenerate conics. This proves the first assertion.

Now take $U = W_o$, $V_1 = K \cap U - \mathrm{Sing}(K)$, $V_2 = \Gamma \cap U - \mathrm{Sing}(\Gamma)$. The theorem implies that for $\sigma \in G$ generic $\sigma(V_1) \cap V_2$ is a non-singular 0-dimensional variety. This implies that $\sigma(K) \cap \Gamma$ is finite (for $\sigma \in G$ generic). To end the proof it is enough to show that $\sigma(K) \cap \Gamma \cap U \subseteq \sigma(V_1) \cap V_2$, again for σ generic. But this is seen applying Kleiman's theorem to the pairs $(\mathrm{Sing}(K) \cap U,\ \Gamma \cap U)$, $(K \cap U,\ \mathrm{Sing}(\Gamma) \cap U)$, where it says that for $\sigma \in G$ generic $\mathrm{Sing}(\sigma(K)) \cap \Gamma \cap U = \emptyset$ and $\sigma(K) \cap \mathrm{Sing}(\Gamma) \cap U = \emptyset$ and the claim follows. □

§ 8. Local characters of a condition

Let K be a condition on conics. We are going to define "local characteristic numbers" of K. In case $K \not\supseteq B$, these numbers are defined to be zero. So assume $K \supseteq B$ and set $\mathcal{O} = \mathcal{O}_{W,B}$, the local ring of W along B, and $m = m_{W,B}$ its maximal ideal. Let X, Y be the rational functions of W defined in §4. Then $m = (X,Y)$, by propositions 4.1 and 4.1ˇ. Let $f \in (X,Y)$ be a local equation of K, i.e., (f) is the ideal $I(K)$ of K in $\mathcal{O}$.

Let $\hat{\mathcal{O}}$ be the m-completion of $\mathcal{O}$. Then there exists a Cohen subfield k_o of $\mathcal{O}$ such that $k_o \xrightarrow{\sim} \hat{\mathcal{O}}/\hat{m} = \mathcal{O}/m = \mathbb{C}(B)$, the field of rational functions of **B**, and such that the inclusion $k_o[X,Y] \subseteq \mathcal{O}$ induces an isomorphism $k_o[X,Y] \xrightarrow{\sim} \hat{\mathcal{O}}$. Consequently we may write f as a formal power series

$$f = \sum_{i,j\geqslant 1} b_{ij}X^iY^j$$

with coefficients $b_{ij} \in k_o$. Associated to this power series we may consider the (Newton-Cramer) set (or diagram) of points $P_{ij} = (i,j) \in \mathbb{R}^2$ such that $b_{ij} \neq 0$. Since f is not divisible by X or Y (because K is degeneration free), this set contains points on each coordinate axis. As usual, for the Newton-Cramer set thus defined we may consider its convex envelope E and the associated Newton-Cramer polygon. The sides of this polygon are the maximal segments contained in the boundary of $E+(\mathbb{R}^+)^2$.

8.1. Definition

A pair of positive integers (p,q) is said to be a pair of *local characteristic numbers* of K if p and q are coprime and there is a side S in the Newton-Cramer polygon whose slope is $-q/p$. The multiplicity of the pair (p,q) is defined as the positive integer r such that r+1 is the number of points on S that have integer coordinates. Symbollically we are going to write r(p,q) to denote that (p,q) is a pair of local characteristic numbers of K with multiplicity r.

It is easy to check that the Newton-Cramer polygon of K, and hence the local characteristic numbers, does not depend on the selection of the local equation f for K, of the local equations X and Y for **A** and **Ǎ**, nor of the Cohen field k_o. In fact the local characteristic numbers of K are related to the structure of the singularity of K along **B**. More information on this aspect can be found in [C.1] or [C.2]. In any event, a proof in a broader context is given in § 18.

§ 9. Halphen's first formula

9.1. Definition

Let Γ be a system and K a condition such that only finitely many conics in Γ satisfy K. Given a conic C, set $e = i_C(K.\Gamma)$ and $e' = i_C(\sigma(K).\Gamma)$, $\sigma \in G$ generic. Then e' will be called *improper intersection multiplicity* of K and Γ at C. For e' to be positive it is necesary and sufficient that C is in $\sigma(K) \cap \Gamma$ for all $\sigma \in G$, that is to say, that C be a point in Γ that improperly satisfies K, and hence, by 6.5, that $C \in B$ and $K \supset B$.

The difference $e-e'$, which is non-negative, will be called *proper intersection multiplicity* of K and Γ at C. We will write $p_C(K.\Gamma)$ to denote it. Thus we have that $p_C(K.\Gamma) = i_C(K.\Gamma)$ unless $C \in B$ and $K \supset B$, in which case $0 \leqslant p_C(K.\Gamma) < i_C(K.\Gamma)$. However, even in this case, $p_C(K.\Gamma)$ may be positive (see example 9.3).

The sum of all proper multiplicities will be denoted $p(K.\Gamma)$, and the sum of all improper multiplicities $imp(K.\Gamma)$. Clearly $p(K.\Gamma)+imp(K.\Gamma) = (K.\Gamma)$, the total intersection number.

The goal in this section is to give an expression of $p(K.\Gamma)$ in terms of the characteristic numbers of K and Γ. This is done in the theorem that follows.

9.2. Theorem (Halphen's first formula)

Let Γ be a system of conics and let (μ,ν) be its global characteristic numbers, $\gamma_1,\ldots,\gamma_h$ its degenerations, and m_i,n_i the order and class of γ_i, $i=1,\ldots,h$. Let K be a condition with global characteristic numbers (α,β) and local characteristic numbers $r_j(p_j,q_j)$, $j=1,\ldots,h'$. Assume that only finitely many conics of Γ satisfy K (i.e., that K does not contain any component of Γ). Then

$$p(K.\Gamma) = \alpha\mu+\beta\nu - \sum_{i,j} r_j \min(m_i q_j, n_i p_j) .$$

Remarks

(a) Given K and Γ there is, by 7.2, a non-empty open set $G' \subset G$ such that $\sigma(K) \cap \Gamma$ is finite for any $\sigma \in G$. Theorem 9.2 then applies to $\sigma(K)$ and Γ, so all conics in $\sigma(K) \cap (\Gamma - B)$ are non-degenerate and, if K is reduced, counted once in the intersection. Furthermore, by the definition of improper intersection number the open set G' may be chosen in such a way that $p_C(\sigma(K).\Gamma) = 0$ for every $C \in \Gamma \cap B$. It follows that $p(\sigma(K).\Gamma)$ is the number of distinct non-degenerate conics which satisfy $\sigma(K)$, for any $\sigma \in G'$, i.e., the formula gives the number of distinct non-degenerate conics in Γ which properly satisfy a reduced condition whose datum is generically chosen.

(b) Chasles' formula gives the number of proper solutions either if Γ does not have degenerations of type B or if $B \notin K$. Otherwise it gives a number greater than Halphen's.

Proof of 9.2.

By the definitions $p(K.\Gamma) = (K.\Gamma) - \mathrm{imp}(K.\Gamma) = \alpha\mu + \beta\nu - \mathrm{imp}(K.\Gamma)$, the latter by Chasles' formula. Now since a branch γ of Γ centered at C contributes to $\mathrm{imp}(K.\Gamma)$ as $i_C(\sigma(K).\Gamma)$, $\sigma \in G$ generic, it is enough to prove that

$$i_C(\sigma(K).\gamma) = \sum_j r_j \min(mq_j, np_j),$$

where m and n are the order and class of γ. This will be done by actually computing $i_{\sigma(C)}(K.\sigma(\gamma))$ for $\sigma \in G$ generic, which is clearly equal to $i_C(\sigma(K).\gamma)$ for $\sigma \in G$ generic.

Using the notations explained in § 8, we see that the coefficients $b_{ij} \in k_o \subset \hat{\mathcal{O}}$ in the power series expansion of f can be approximated, up to any preassigned order, by elements of $\mathcal{O}$. Thus, if we let s be an integer such that $s \geq (K.\Gamma)$ and $s \geq i+j$ for any point (i,j) in the Newton-Cramer polygon of f, then there exist elements $b'_{ij} \in \mathcal{O}$ such that $b'_{ij}-b_{ij} \in m^{s+1}\hat{\mathcal{O}}$, so that

$$\sum_{\substack{i,j \geq 1\\ i+j \leq s}} b'_{ij}X^iY^j - f \in m^{s+1}\hat{\mathcal{O}} \cap \mathcal{O} = m^{s+1}$$

(we take $b'_{ij} = 0$ if $b_{ij} = 0$). Since b_{ij} is invertible if it is non-zero, it turns out that b'_{ij} is also invertible when $b_{ij} \neq 0$, and so the Newton-Cramer polygon associated to $\sum b'_{ij}X^iY^j$ coincides with the Newton-Cramer polygon of f.

Let $B_o \subseteq B$ be a non-empty open set of B such that

(a) The functions X,Y,f, and b'_{ij} are regular on B_o;

(b) For each $C \in B_o$, X, Y, and f respectively generate, in $\mathcal{O}_{W,C}$, the ideals of $\mathbf{A}$, $\check{\mathbf{A}}$, and K; and

(c) b'_{ij} is either identically zero or everywhere non-zero on B_o (this can be accomplished because $b'_{ij} \in \mathcal{O}^*$ if $b_{ij} \neq 0$).

In the computation of $i_{\sigma(C)}(K.\sigma(\gamma))$, $\sigma \in G$ generic, the conic $D = \sigma(C)$ belongs to B_o. Let $\mathcal{O}_D = \mathcal{O}_{W,D}$, so that $\mathcal{O}_D \subseteq \mathcal{O}$ and $m \cap \mathcal{O}_D \subseteq m_D$, the maximal ideal of $\mathcal{O}_D$. Since X,Y,f, $b'_{ij} \in \mathcal{O}_D$, we actually have that

$$\sum_{\substack{i,j \geq 1\\ i+j \leq s}} b'_{ij}X^iY^j - f \in m_D^{s+1},$$

and since $i_D(K.\sigma(\gamma)) = \mathrm{ord}_{\sigma(\gamma)}(f) \leq (K.\Gamma) \leq s$, it turns out that

$$i_D(K.\sigma(\gamma)) = \mathrm{ord}_{\sigma(\gamma)}(f'), \qquad \text{where} \qquad f' = \sum_{i+j \leq s} b'_{ij}X^iY^j .$$

Let t be a parameter for the branch $\sigma(\gamma)$, and set $\tilde{\xi} = \tilde{\xi}(t)$ to denote the restriction of an element $\xi \in \mathcal{O}_D$ to $\sigma(\gamma)$. Then $\tilde{X}$ and $\tilde{Y}$ are power series in t of orders n and m, respectively. Hence we may assume that

$$\tilde{X} = t^n + \dots , \qquad \tilde{Y} = ct^m + \dots ,$$

where $c \neq 0$ and where ... stands for higher order terms.

We will set $c = c(\sigma(\gamma))$ if we need to stress that c depends on $\sigma(\gamma)$. Then the contribution in $\tilde{f}'$ of a monomial $b'_{ij}X^iY^j$, $b'_{ij} \neq 0$, is a power series in t of the form

$$b'_{ij}(C)\ c^j\ t^{ni+mj} + \dots$$

Now our task will be first to compute $\ell = \min(ni+mj)$, for all i,j such that $b_{ij} \neq 0$, and then show that the coefficient of t^ℓ in $\tilde{f}'$ is non-zero, so that we will have $i_D(K.\sigma(\gamma)) = \mathrm{ord}_t \tilde{f}' = \ell$.

For the computation of ℓ, notice that we only need take into account the points (i,j) on the Newton-Cramer polygon. We will assume that the sides of this polygon are indexed successively starting with the side which has a vertex (say $(a,0)$) on the X-axis. With this convention a point on the i-th side of the polygon can be expressed as

$$(a,0) + r_1(-p_1,q_1) + \dots + r_{i-1}(-p_{i-1},q_{i-1}) + r(-p_i,q_i),$$

$0 \leq r \leq r_i$. This point thas integer coordinates when r itself is an integer. The exponent of t corresponding to such a point is

$$\begin{aligned}\ell' &= (a - r_1p_1 - \dots - rp_i)n + (r_1q_1 + \dots + rq_i)m = \\ &= an + r_1(-p_1n+q_1m) + \dots + r_{i-1}(-p_{i-1}n+q_{i-1}m)+r(-p_in+q_im).\end{aligned}$$

Because of the way the sides of the polygon have been indexed we have that the quotients q_i/p_i increase with i, so that there exists j such that:

$$q_j/p_j < n/m \leqslant q_{j+1}/p_{j+1}$$

(we set $j=0$ if $n/m \leqslant q_1/p_1$, and $j=h'$ if $q_{h'}/p_{h'} < n/m$. Then we see that

$$-p_i n + q_i m \begin{cases} < 0 & \text{for } i \leqslant j \\ \geqslant 0 & \text{for } i > j \end{cases}$$

and hence ℓ' will be minimum only when $i-1=j$ and $r=0$ if $n/m < q_{j+1}/p_{j+1}$ (i.e., the minimum is taken exactly at a vertex of the polygon), or when $i-1=j$ and for any r such that $0 \leqslant r \leqslant r_{j+1}$ if $n/m = q_{j+1}/p_{j+1}$ (i.e., the minimum is taken at all points that lie on the $(j+1)$-th side of the polygon). In any case the minimum we are seeking is

$$\begin{aligned}\ell &= (r_1q_1+\ldots+r_jq_j)m + (a-r_1p_1-\ldots-r_jp_j)n = \\ &= (r_1q_1+\ldots+r_jq_j)m + (r_{j+1}p_{j+1}+\ldots+r_{h'}\cdot p_{h'})n = \\ &= \sum r_i \min(mq_i, np_i) .\end{aligned}$$

The proof of the theorem will be completed if we show that the coefficient of t^ℓ in $\tilde{f}$ is non-zero. Notice that this is automatically true if $n/m < q_{j+1}/p_{j+1}$. If $n/m = q_{j+1}/p_{j+1}$, then the coefficient of t^ℓ has the form $g(c)$, where g is a non-zero polynomial which only depends on γ, and not on σ. Since g has only a finite number of roots, it will be enough to show that there are elements $\tau \in G$ leaving D fixed and such that $c(\tau\sigma(\gamma))$ is not a root of g. This will be done by seeing that $c(\tau\sigma(\gamma))$ takes infinitely many values as a function of τ, τ leaving D fixed.

Indeed, let P and u be the point and the line used to define X and Y (see §4). Let τ be the homology with center at P, axis the line v such that v^2 is the point component of D, and modulus $z \in \mathbb{C}^*$. Then it is clear that $\tau(D) = D$. Since τ leaves invariant the lines through P, $X \circ \tau = X$. However, $Y \circ \tau = zY$, as one sees considering the restriction of τ to u. Thus, for such a τ, $c(\tau\sigma(\gamma)) = zc(\sigma(\gamma))$. □

9.3. Example

We construct a condition K, a system Γ and a conic C such that $C \in B$, $K \supset B$, and $p_C(K \cdot \Gamma) = 1$.

Let P_o, P_1, P_2, U be a projective system of coordinates. Let $U'=(1,1,0)$ and $Q_t=(1,0,t)$, $t \neq 0$. Now let Γ be the pencil of conics

$$C_\lambda: \quad \lambda(X_o^2-2X_1X_2) + X_2^2$$

and let K be the condition defined by

$$K_t: \quad Y = Y^2+X .$$

Here Y, X are defined as follows:

$$Y = Y(C) = [(P_2,P_o,Q_t,R)-(P_2,P_o,Q_t,R')]^2 \quad ,$$

where R, R' are the points where C intersects the line P_oP_2, and

$$X = X(C) = [(u_1,u_o,u',v)-(u_1,u_o,u',v')]^2 \quad ,$$

where v, v' are the tangents to C drawn from P_2 and $u_1=P_2P_o$, $u_o=P_2P_1$, and $u'=P_2U'$.

The system Γ has the conic $X_2^2=0$, with P_1 as a double focus, as a degeneration of type B. Let C denote this conic. Since $Y(C_\lambda) = 4\lambda/t^2$ and $X(C_\lambda) = 4\lambda$, as one easily computes, then $i_C(K \cdot \Gamma)$ is the multiplicity in λ of the polynomial

$$4t^{-4}\lambda(4\lambda+t^4-t^2).$$

For $t= \pm 1$, this multiplicty is 1, while for $t=\pm 1$ it is equal to 2. Since clearly all the conditions K_t are in the same orbit we see that for $t=\pm 1$ $imp_C(K_t \cdot \Gamma) = 1$ and $p_C(K_t \cdot \Gamma) = 1$. □

§ 10. Computation of characteristic numbers

In this section we establish a method for determining the (local and global) characteristic numbers of a condition. This is done by intersecting the condition with a suitable surface and by relating the characteristic numbers to numerical characters of the intersection curve.

10.1. Lemma

Given a condition K, there exists a non-empty open set $B_o \subseteq B$ such that if S is an irreducible surface cutting B transversally at $C \in B_o$ then the restrictions $\bar{X}$ and $\bar{Y}$ of X and Y to S, respectively, form a system of parameters for S in a neighbourhood of C. Furthermore, if $S \nsubseteq K$ then the Newton-Cramer polygon of the curve $K \cap S$ with respect to the parameters $\bar{X},\bar{Y}$ coincides with the Newton-Cramer polygon of K.

Proof

With the same notations as in the proof of 9.2, given $C \in B_o$ there exists $b'_{ij} \in \mathcal{O}_C$ such that the Newton-Cramer polygon of

$$\sum_{i+j \leqslant s} b'_{ij} X^i Y^j$$

coincides with the Newton-Cramer polygon of K. Moreover,

$$f - \sum_{i+j \leqslant s} b'_{ij} X^i Y^j \in m_C^{s+1}$$

and $b'_{ij} \neq 0$ if $b_{ij} \neq 0$. Now the transversality condition implies that $\bar{X},\bar{Y}$ generate the maximal ideal $\bar{m}_C$ of $\mathcal{O}_{S,C}$, and so we have

$$\bar{f} - \sum_{i+j \leqslant s} \bar{b}'_{ij} \bar{X}^i \bar{Y}^j \in \bar{m}_C^{s+1} \quad ,$$

where for an element $h \in \mathcal{O}_C$, $\bar{h}$ denotes its restriction to S. From this relation the last part of the statement follows immediately. □

10.2. Corollary

The local characteristic numbers of degeneration free conditions are additive.

Proof

Lemma 10.1 allows us to deal with locally plane curves instead of conditions (not necessarily reduced). Now for locally plane curves it is well known (cf. [Wal]) that the negative slopes of the sides of the Newton-Cramer polygon are the first exponents in the Puiseux series of the branches of the curve and that the sum of the orders of the branches corresponding to a given side is the number of integral points on that side decreased by 1 from what the additivity for curves is clear.

To describe the particular surface S which will be used, fix a triangle T in $\mathbb{P}_2$ and set x_i, u_i, $i=0,1,2$, to denote its vertexes and sides, respectively, with u_i the side opposite x_i. Let $S=S_T$ be the surface in $\mathbf{W}$ of all conics for which T is a self-polar triangle. If we take T as a projective system of coordinates then the conics $C=(c,c')$ of S_T are precisely those for which the matrices of c and c' are diagonal.

Let us denote by u_i^2 the point conic consisting of the line u_i counted twice and by x_i^2 the envelope consisting of the pencil of lines through x_i counted twice. Then $p(S_T)$ is a plane in $\mathbb{P}_5$ and $p\colon S_T \longrightarrow p(S_T)$ is the blowing up of $p(S_T)$ at the points u_o^2, u_1^2, and u_2^2. Similarly, $t(S_T)$ is a plane in $\check{\mathbb{P}}_5$ and $t\colon S_T \longrightarrow t(S_T)$ is the blowing up of $t(S_T)$ at the points x_o^2, x_1^2, and x_2^2. Equivalently, S_T is the graph of the quadratic Cremona transformation $p(S_T) \longrightarrow t(S_T)$ whose fundamental triangles are (u_o^2,u_1^2,u_2^2) in $p(S_T)$ and (x_o^2,x_1^2,x_2^2) in $t(S_T)$.

The intersection $S_T \cap \mathbf{A}$ consists of the three mutually disjoint pencils H_i, $i=0,1,2$, defined as follows: H_i is the pencil of pairs of lines that have x_i as a double point and harmonically divide the pair of lines $(u_j,u_{j'})$

going through x_i. Each of these pencils has multiplicity one in the intersection — notice that $p(\mathbf{A}) = D$ and $p(S_T)$ are respectively a cubic hypersurface and a plane in $\mathbb{P}_5$.

Dually, the intersection $S_T \cap \check{\mathbf{A}}$ consists of three disjoint ranges $\check{H}_i$, $i=0,1,2$, where the conics in $\check{H}_i$ have u_i^2 as point component and two points on u_i harmonically dividing the two vertexes of T on u_i as line component. The multiplicity of $\check{H}_i$ in $S_T \cap \check{\mathbf{A}}$ is one.

Each pencil H_i meets precisely two $\check{H}_j$'s, and conversely, each $\check{H}_i$ meets two H_j's. In fact if (i,j,k) is a permutation of $(1,2,3)$, then H_i intersects $\check{H}_j$ and $\check{H}_k$ but not $\check{H}_i$, and $\check{H}_i$ intersects H_j and H_k, but not H_i. If for $i \neq j$ we let C_{ij} denote the conic (u_i^2, x_j^2), then $\check{H}_i \cap H_j = \{C_{ij}\}$. From this it follows that $S_T \cap \mathbf{B}$ consists of the six conics C_{ij} and each of them has multiplicity one in this intersection.

Let K be a reduced (degeneration free) condition. Let (α,β) and $r_i(p_i,q_i)$, $i=1,\dots,s$, denote the global and local characteristic numbers of K, respectively. Given a triangle $T = \{x_o,x_1,x_2\}$ in $\mathbb{P}_2$ we will as before denote by u_o,u_1,u_2 the sides of T, with the convention that $x_i \notin u_i$. We choose such a triangle T in such a way that the conics $C_{ij} = \{u_i^2, x_i^2\}$ belong to the open set B_o of lemma 10.1 and that $S_T \cap K$ is a reduced curve. Let $\mathcal{T}$ be the triangle of $\mathbb{P}_5$ whose vertexes are u_o^2, u_1^2, u_2^2, so that the plane that $\mathcal{T}$ determines consists of the point conics which have T as a self-polar triangle, i.e., $\mathcal{T}=p(S_T)$. With these notations we have:

10.3. Theorem

The conic locus componenets of the conics in S_T that satisfy K form a curve K' in the plane of $\mathcal{T}$ which has the following properties:

(a) $\deg K' = \alpha+2\beta$;

(b) The vertexes of $\mathcal{T}$ are points of multiplicity β on K'; and

(c) Given a vertex of $\mathcal{T}$ and a side concurring in it, for each branch

γ of K' with center at the vertex and tangent to the given side there exist a unique i such that $\text{ord}\,\gamma = \ell p_i$, class $\gamma = \ell q_i$, ℓ a positive integer.

Moreover, $\sum \ell = r_i$, where the sum ranges over all such branches with fixed i.

Proof

Let $S = S_T \subset \mathbf{W}$ be the surface of conics which have T as a self-polar triangle and set $\bar{K} = K \cap S$. We will analyze the behaviour of $\bar{K}$ in the neighbourhood of C_{o1}. If $\bar{X}$ and $\bar{Y}$ denote the restrictions of X and Y to S (notations of lemma 10.1), then $\bar{X}$ is a local equation of $\check{H}_1$ and $\bar{Y}$ of H_o, both at C_{o1}. By lemma 10.1, for each branch $\bar{\gamma}$ of $\bar{K}$ at C_{o1} there exists a unique i, $1 \leqslant i \leqslant s$, such that the parametric equations of $\bar{\gamma}$ are of the form

$$\bar{X} = \lambda^{\ell p_i} f(\lambda), \qquad \bar{Y} = \lambda^{\ell q_i} g(\lambda), \qquad f(0) \neq 0, \qquad g(0) \neq 0 \quad .$$

Moreover, $\sum \ell = r_i$, where the sum ranges to all such branches $\bar{\gamma}$ with i fixed. On the other hand, by Chasles' theorem 5.5 the intersection number of $\bar{K}$ with H_1 and $\check{H}_o$ are, respectively, α and β. Now $K' = p(\bar{K})$, $p(\check{H}_o) = u_o^2$, and since p is, locally at u_o^2, the blowing up of u_o^2, we see that K' has multiplicity β at u_o^2. Furthermore, the correspondence $\bar{\gamma} \longmapsto \gamma = p(\bar{\gamma})$ establishes a bijection between the branches $\bar{\gamma}$ of $\bar{K}$ at C_{o1} and the branches of K' at u_o^2 which are tangent to $p(H_1)$, one of the sides of $\mathcal{T}$ going through u_o^2 (the other side is $p(H_2)$). Now if $\bar{\gamma}$ is the branch given by the equations above, then γ is a branch of order ℓp_i and its intersection multiplicity with $p(H_1)$ is $\ell(p_i + q_i)$, so that its class is ℓq_i. Moreover, the order of K' is $\alpha + 2\beta$, because $K' = p(K) \cap p(S_T)$ is a plane section of $p(K)$. The proof of the theorem is complete if we take into account that the same analysis that we have applied to C_{o1} works for any C_{ij}, $i \neq j$. □

10.4. Corollary

If K is a degeneration free condition with global characteristic numbers (α,β) and local characteristic numbers $r_i(p_i,q_i)$, $i=1,\ldots,s$, then

$$\alpha \geqslant 2\sum r_i q_i, \quad \beta \geqslant 2\sum r_i p_i.$$

Proof

By 10.2 we may assume that K is irreducible. With the same notations as in the theorem, notice that the intersection number of $p(H_1)$ with K' is $\alpha+2\beta$, and hence

$$\alpha+2\beta \geqslant i_{u_o^2}(p(H_1).K') + i_{u_2^2}(p(H_1).K') .$$

But by the theorem both summands are $= \beta + \sum r_i q_i$. The second inequality can be seen either by duality or else by noticing that since β is the multiplicity of K' at the vertex it cannot be greater than the sum of the orders of the branches of K' centered at u_o^2. □

§ 11. Examples

11.1. (The conditions $S_{p,q}$)

Let X and Y be functions defined as at the end of §4, say by means of a line u and a point P, $P \in u$. For any non-zero polynomial $Q(X,Y)$, the relation $Q(X,Y) = 0$ determines a hypersurface in the open set of conics that do not pass through P and are not tangent to u. We will set K_Q to denote the clousure of such a hypersurface.

To get some intuitive feeling for this type of hypersurfaces recall that in Euclidean terms the function X is the square of the lenghth of the segment cut out on u by a conic and that Y is the square of the tangent of the angle between the tangents to a conic drawn from P.

The local characteristic numbers of K_Q can be obtained directly from the Newton–Cramer polygon of Q, inasmuch as $Q(X,Y)$ is a local equation of K_Q at the generic point of B.

In particular, set $S_{p,q}$ to denote the condition K_Q obtained taking $Q(X,Y) = \lambda X^p - Y^q$, where p,q are coprime positive integers and λ is a non-zero complex number. Then it is clear that $S_{p,q}$ has a single pair of local characteristic numbers, namely (p,q). Moreover, its global characteristic numbers are given by $\alpha = 2q$, $\beta = 2p$. This can be seen directly using the description of X and Y in § 4, or by means of the method described in § 10.

In fact if P_o, P_1, P_2, Q is a projective system of coordinates and u_o, u_1, u_2, v is the dual system, let Q_2 be the projection of Q on u_2 from P_2, and let v_o be the line constructed as Q_2 but dually, using P_o and u_o (see figure).

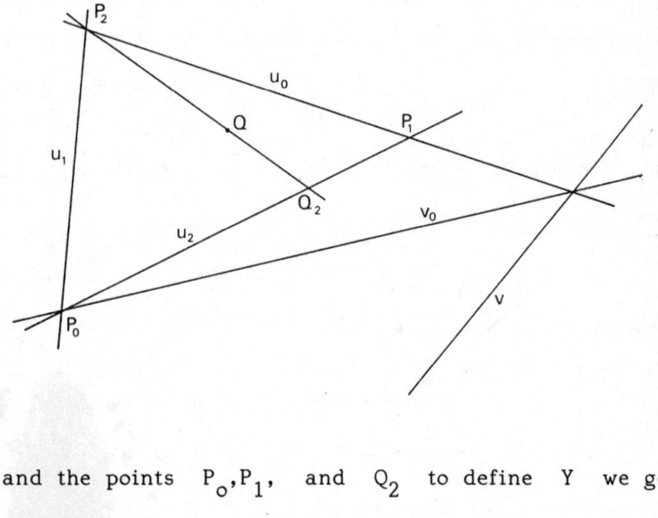

Then if we use u_2 and the points P_o, P_1, and Q_2 to define Y we get

$$Y(C) = 4\,\bar{a}_{22}/a_{oo}^2$$

(see the proof of 4.3). If we define X dually, with respect to P_o, u_2, u_1, v_o then

$$X(C) = 4\,\bar{\alpha}_{oo}/\alpha_{22}^{2}$$

From these expressions it is easy to see that the hypersurface $S_{p,q}$ defined by means of X and Y is given by the bihomogeneous equation

$$\lambda\,\bar{a}_{22}^{q}\alpha_{22}^{2p} = a_{oo}^{2q}\bar{\alpha}_{oo}^{p} \quad .$$

This shows that $S_{p,q}$ lies in the linear pencil generated by the (linearly equivalent) divisors $q\check{A}+(2p+q)L_{u_2}$ and $pA+(p+2q)L_{P_o}$. In particular the global characteristic numbers of $S_{p,q}$ are $(2q,2p)$.

11.2. Proposition

The conditions

$$\alpha \geqslant 2\sum_{i=1}^{s} r_i q_i, \qquad \beta \geqslant 2\sum_{i=1}^{s} r_i p_i$$

are necessary and sufficient in order that it exists a degeneration free condition whose global and local characteristic numbers are, respectively, (α,β) and $r_i(p_i,q_i)$, $i=1,\ldots,s$.

Proof

The conditions are necessary by corollary 10.4. To see that they are sufficient, let

$$K = \sum_{i=1}^{s} r_i S_{p_i,q_i} + (\alpha - 2\sum_{i=1}^{s} r_i q_i)L + (\beta - 2\sum_{i=1}^{s} r_i q_i)\check{L}$$

Then K is effective and degeneration free, and by corollary 10.2 its global and local characteristic numbers are, respectively, (α,β) and $r_i(p_i,q_i)$, $i=1,\ldots,s$. □

We proceed with the examples.

11.3. (Conditions expressing a relation to a fixed conic)

Let C and C' be conics with matrices $((a_{ij}),(\alpha_{ij}))$ and $((b_{ij}),(\beta_{ij}))$, respectively. As usual we will denote by Δ, Θ, Θ', and Δ' the fundamental invariants of C and C', so that

$$\det(\lambda(a_{ij}) + \mu(b_{ij})) = \Delta\lambda^3 + \Theta\lambda^2\mu + \Theta'\lambda\mu^2 + \Delta'\mu^3.$$

In the special case that (a_{ij}) is a diagonal matrix, say $\mathrm{diag}(a_o,a_1,a_2)$ (so that the triangle of reference T is self-polar with respect to C),

$$(*)\qquad \begin{aligned} \Delta &= a_o a_1 a_2 \\ \Theta &= b_{oo}a_1a_2 + b_{11}a_oa_2 + b_{22}a_oa_1 \\ \Theta' &= a_o\beta_{oo} + a_1\beta_{11} + a_2\beta_{22} \\ \Delta' &= \det(b_{ij}) \end{aligned}$$

These expressions are useful because if K is the condition that a conic C satisfies a relation to a fix conic C' expressed as a polynomial relation

$$(**)\qquad Q(\Delta,\Theta,\Theta',\Delta') = 0$$

between the fundamental invariants of the pair C,C', then the equation of the curve $p(K\cap S_T)$ is obtained substituting $\Delta,\Theta,\Theta',\Delta'$ in (**) for the expressions (*).

Here are a few examples (cf. [S-K], Ch. VIII, 3):

(K_1) The condition that a conic C is triangularly circumscribed to a conic C' is expressed by the relation

$$\Theta'^2 - 4\Delta'\Theta = 0.$$

After performing the above substitution one gets a conic in the plane $p(S_T)$ that does not go through the vertexes of $\mathcal{T}$ (notations as in theorem 10.3). Therefore $\alpha=2$, $\beta=0$ and there are no local characteristic numbers.

($\check{K}_1$) By duality the condition that a conic C is triangularly inscribed in a conic C' will have $\alpha=0$, $\beta=2$, and no local characteristic numbers. Notice however that such a condition is expressed by the relation

$$\Theta^2 - 4\Delta\Theta' = 0 ,$$

which, after performing the substitution (*), yields a *quartic* curve in $p(S_T)$. This quartic has an ordinary node at each vertex of $\mathcal{T}$ the tangents of which are distinct from the sides of $\mathcal{T}$ concurring thereto. Thus we get indeed $\alpha=0$, $\beta=2$, and that there are no local characteristic numbers.

(K_2) The condition that a conic C has a pair of common tangents with C' whose intersection point is collinear with two points of $C \cap C'$ is expressed by the (self-dual) relation

$$\Delta\Theta'^3 + \Delta'\Theta^3 = 0 .$$

After substituting (*) in it one gets a sextic in $p(S_T)$ with an ordinary node at each vertex whose tangents coincide with the sides of $\mathcal{T}$ concurring thereto. Therefore $\alpha=\beta=2$ and $(1,1)$ is the only pair of local characteristic numbers.

11.4. (Tangency to a curve)

Let F be an irreducible curve in $\mathbb{P}_2$. Given a point $P \in F$ and a conic C through P, we will say that C is tangent to F at P if F and C have a common tangent at P. Let $K = K_F$ be the set of conics that are tangent to F.

Notice that if we had taken the tangency of C and F at P to mean that $i_P(C.F) > 1$ then the set of conics tangent to F (in this weaker sense) would be $K_F \cup K'$, where K' is the set of conics going through a singular point of F.

Let g be the geometric genus of F, d its degree and $\check{d}$ its class. Let d_i, $1 \leq i \leq r$, be the orders of the singular branches γ_i of F (i.e., branches whose order is greater than one). Fix a generic pencil L of conics and consider the map

$$\sigma : F \longrightarrow L$$

such that $\sigma(P)$ is the unique conic in L going through P. Clearly $\deg(\sigma) = 2d$ and so, by Hurwitz's formula,

$$2g - 2 = -4d + \deg(R) ,$$

where R is the ramification divisor of σ. Now since L is generic, the finitely many conics in L that are tangent to F have a single tangency. It turns out that R is the sum of the points P on F for which there is a conic of L which is tangent to F at P plus the singular branches γ_i of F, the i-th such being counted with multiplicity d_i-1. Therefore

$$\deg R = \deg p(K_F) + \sum_{r=1}^{r} (d_i-1) = \alpha + 2\beta + \sum_{i=1}^{r} (d_i-1) ,$$

where (α, β) are the global characteristic number of K_F. Comparing with the previous relation we get that

$$\alpha + 2\beta = 2g - 2 + 4d - \sum (d_i-1) .$$

On the other hand

$$2g - 2 + 2d = \check{d} + \sum (d_i-1) ,$$

as is seen by a similar argument with a generic pencil of lines instead of L. Hence

$$\alpha + 2\beta = 2d + \check{d} .$$

Dually,

$$2\alpha + \beta = d + 2\check{d} ,$$

and therefore $\alpha=\check{d}$, $\beta=d$. Since $B \notin K_F$, there are no local characteristic numbers.

§ 12. Strict equivalence of conditions

Let Z be the group of divisors that do not have **A** or $\check{\mathbf{A}}$ as components. The elements of Z will be called *degeneration free divisors* of **W**. The effective divisors in Z is what we have been calling degeneration free conditions. The additive character of Halphen's formula 9.2 implies that it is also valid for degeneration free divisors if the notions of local and global characteristic numbers are extended to Z by additivity.

12.1. Definition (Halphen [H2], § 27)

Given $D_1, D_2 \in Z$, D_1 is said to be *strictly equivalent* to D_2 iff for any system of conics Γ the proper intersection numbers of Γ with D_1 and D_2 coincide whenever they are defined.

Halphen calls this relation "equivalence de conditions". It is a numerical equivalence, but relative to the number of proper solutions. As we will see in a moment, this is a much stronger equivalence than the usual numerical equivalence.

The quotient group of Z under the strict equivalence will be called Halphen's group and denoted by $\mathrm{Hal}(\mathbf{W})$. The strict equivalence class of a divisor $D \in Z$ will be denoted by $<Z>$.

12.2. Theorem

Two degeneration free divisors are strictly equivalent iff their global and local characteristic numbers coincide.

Proof

That the condition is sufficient is clear due to the extension of Halphen's formula to degeneration free divisors. To see the necessity, it is enough to consider the case of two degeneration free conditions K and K'. Let α, β, and $r_i(p_i, q_i)$, $1 \leq i \leq s$, be the global and local characteristic numbers of K, and α', β', and $r'_j(p'_j, q'_j)$, $1 \leq j \leq s'$ the like characters for K'. Assume that K and K' are strictly equivalent. We want to show that the two sets of characteristic numbers are equal.

That the global characters coincide can be seen by taking for Γ a pencil or a range of conics, cases in which Halphen's formula does not involve local terms, and expressing that $p(K.\Gamma) = p(K'.\Gamma)$.

Now it is easy to see that for any pair (m,n) of positive integers there exists a system Γ of conics not contained in any component of K or K' and such that it has a unique degeneration γ of type B whose order and class are m and n, respectively(*). The definition of strict equivalence between K and K', together with the equality of the global characteristic numbers of K and K', yield the equality

$$(*) \qquad \sum_{i=1}^{s} r_i \min(mp_i, nq_i) = \sum_{j=1}^{s'} r'_j \min(mp'_j, nq'_j) ,$$

which must be fullfilled for all pairs (m,n) of positive integers.

Without loss of generality we may assume that

$$p_i/q_i < p_{i+1}/q_{i+1} \quad \text{and} \quad p'_j/q'_j < p'_{j+1}/q'_{j+1}, \quad 1 \leq i < s,\ 1 \leq j < s'.$$

If we had that $r_i = r'_i$ and $p_i/q_i = p'_i/q'_i$ for $1 \leq i \leq \inf(s,s')$ then (*) would imply immediately that $s=s'$ and hence the two sets of characteristic

(*) Such a system is actually constructed in § 19

numbers would coincide. So assume that for some i, $1 \leq i < \inf(s,s')$, $r_h = r'_h$ and $p_h/q_h = p'_h/q'_h$ for all $h < i$, but that either

$$p_i/q_i \neq p'_i/q'_i$$

or else

$$p_i/q_i = p'_i/q'_i \quad \text{and} \quad r_i \neq r'_i \quad .$$

From this we will derive a contradiction and hence the theorem will be proved.

Let us first consider the assumption $p_i/q_i \neq p'_i/q'_i$. Again, without loss of generality, we may assume that $p_i/q_i < p'_i/q'_i$. Now choose (m,n) so that

$$(**) \qquad p_i/q_i < n/m < p'_i/q'_i \quad .$$

Then one has the following relations:

$$\begin{array}{lll} p_j m = p'_j m < q_j n = q'_j n & \text{for} & 1 \leq j \leq i-1, \\ p_i m < q_i n, \quad p'_i m > q'_i n & & \\ p_j m > q_j n & \text{for} & i < j \leq s \quad , \\ p'_j m > q'_j n & \text{for} & i < j \leq s' \; . \end{array}$$

Combining these relations with (*) we obtain that

$$r_i p_i \frac{m}{n} = \sum_{j=1}^{s'} r'_j q'_j - \sum_{j=i+1}^{s} r_j q_j \quad ,$$

which contradicts the fact that there are infinitely many rational numbers satisfying (**).

The assumption $p_i/q_i = p'_i/q'_i$, $r_i \neq r'_i$, leads likewise to a contradiction. This time choose (m,n) so that

$$p_i/q_i = p'_i/q'_i < m/n < \min(p_{i+1}/q_{i+1}, \; p'_{i+1}/q'_{i+1}) \; .$$

A similar argument then shows that

$$(r_i-r_i')p_i \frac{m}{n} = \sum_{j=i+1}^{s'} r_j'q_j' - \sum_{j=i+1}^{s} r_jq_j ,$$

which gives a contradiction of the same sort as above.

12.3. Corollary

Hal(**W**) is a free abelian group for which the classes $\langle L\rangle$. $\langle \check{L}\rangle$, and $\langle S_{p,q}\rangle$, where (p,q) is any pair of positive coprime integers, are a free basis. Furthermore, a linear combination of distinct elements of such a basis is the class of a degeneration free condition iff the coefficients of the combination are non-negative and not all zero.

Proof

First notice that given a pair of positive coprime integers (p,q) then the conditions $S_{p,q}$ introduced in 11.1 are strictly equivalent to each other. In fact the characters of $S_{p,q}$ are independent of the choices made in their definition. Similarly, L and $\check{L}$ uniquely define classes under the strict equivalence, since they do not have local characteristic numbers and their global characteristic numbers are $(1,0)$ and $(0,1)$, respectively.

Next notice that given a degeneration free divisor D with characteristic numbers α, β, $r_i(p_i,q_i)$, $1 \leqslant i \leqslant s$, then the divisor

$$D' = (\alpha - 2\sum r_iq_i)L + (\beta - 2\sum r_ip_i)\check{L} + \sum r_iS_{p_i,q_i}$$

has the same characteristic numbers as D, so that $\langle D\rangle = \langle D'\rangle$ is a linear combination with integer coefficients of $\langle L\rangle$, $\langle \check{L}\rangle$, and the $\langle S_{p,q}\rangle$.

If D is a condition, then by 11.2 these coefficients are non-negative and not all zero. Conversely, if they are non-negative and not all zero then D is strictly equivalent to a condition. That $\langle L\rangle$, $\langle \check{L}\rangle$, and the $\langle S_{p,q}\rangle$

are $\mathbb{Z}$-linearly independent is clear enough.

12.4. Remark

The conditions L, $\check{L}$, and $S_{p,q}$ are called *elementary conditions* by Halphen. He proves ([H.2], theorem IV) that they generate $\mathrm{Hal}(\mathbf{W})$. This fact turns out to be a fundamental one in [H.3]. However, it is not until the end of [H.3] that Halphen characterizes in intrinsic terms, for a condition D, the coefficient of a given elementary condition in a linear combination of elementary conditions expressing D. This characterization actually implies that the linear combination that expresses a given D is unique. □

§13. Systems of conics contained in $S_{p,q}$

In this section we are interested in analyzing the possible degenerations of a system of conics under the hypothesis that all of them satisfy an elementary condition.

To start with, let P be a point and u a line in $\mathbb{P}_2$ such that $P \in u$. Let $S_{p,q}$ be a condition defined as in 11.1 using P and u (here we need not specify the other choices in the definition). Then we have:

13.1. Proposition

Let Γ be a system of conics such that $\Gamma \subset S_{p,q}$ and let γ be a degeneration of type B of Γ. Set (v^2,Q^2) to denote the center of γ, so that v is a line and Q a point on it, and let m,n be the order and class of γ, respectively. With these notations,

(a) If $m/n > p/q$, then v goes through P;

(b) If $m/n < p/q$, then Q lies on u;

(c) Therefore, if v does not go through P and Q does not lie on u then

$m/n = p/q$.

Proof

In §4 we saw that

$$\mathrm{div}_{\mathbf{W}}(X) = \mathbf{A} + L_P - 2\check{L}_u$$
$$\mathrm{div}_{\mathbf{W}}(Y) = \check{A} + \check{L}_u - 2L_P \quad ,$$

where X, Y are the functions used to define $S_{p,q}$, $\mathrm{div}_{\mathbf{W}}(f)$ is the divisor on $\mathbf{W}$ of the rational function f, and where L_P and $\check{L}_u$ are, respectively, the hypersurface of conics which go through P and which are tangent to u. Consequently,

$$\mathrm{ord}_\gamma(X) = n + (\gamma . L_P) - 2(\gamma . \check{L}_u)$$
$$\mathrm{ord}_\gamma(Y) = m + (\gamma . \check{L}_u) - 2(\gamma . L_P) \quad ,$$

with the convention that, for a hypersurface H, $(\gamma . H) = \infty$ iff γ is contained in H.

Since $\Gamma \subset S_{p,q}$, we have

$$p \, \mathrm{ord}_\gamma(X) = q \, \mathrm{ord}_\gamma(Y)$$

and so

$$(*) \qquad pn - qm = (2p+q)(\gamma . \check{L}_u) - (p+2q)(\gamma . L_P) \ .$$

Now if we assume that $m/n < p/q$, i.e., $pn-qm > 0$, then necessarily $(\gamma . \check{L}_u) > 0$. But this means that the origin (v^2, Q^2) of γ is tangent to u and hence that $Q \in u$. This proves (b). The proof of (a) is similar. □

The previous proposition can be somewhat extended to other conditions. In order to explain this, (symbollically) denote L_P by $S_{0,1/2}$ and $\check{L}_u$ by $S_{1/2,0}$. Actually we will think of $S_{0,1/2}$ as defined by means of point P and an undetermined line (which we do not need to specify), and of $S_{1/2,0}$ as defined by an undetermined point and the line u. With these conventions notice that proposition 13.1(a) is also true for $S_{0,1/2}$ and $S_{1/2,0}$. Indeed, for $S_{0,1/2}$ we have $p=0$, so that the hypothesis $m/n > p/q$ in 13.1(a) is always true, and obviously the thesis is also true; on the other hand the hypothesis in 13.1(b) is never true, so that 13.1(b) is correct for $S_{0,1/2}$ too. The case of $S_{1/2,0}$ works now by duality.

Proposition 13.1 is also true for the degenerations of Γ of types A and $\check{A}$. In fact for the degeneration of type A we have that $m=0$, $n\neq 0$ so that the hypothesis of 13.1(a) cannot occur, whereas the hypothesis in 13.1(b) is always true, and so is the thesis, because (*), which is still correct for γ, implies that $\gamma.\check{L}_u > 0$ and so the double point Q of the center of γ lies on u. The case of a degeneration of type $\check{A}$ is dual. Sumarizing we have:

13.2 The conclusions of 13.1 hold true in the following cases:

(a) For the condition L_P when considered as a $S_{0,1/2}$ relative to a point P and an unspecified line.

(b) For the condition $\check{L}_u$ when considered as a $S_{1/2,0}$ relative to an unspecified point and to a line u.

(c) For degenerations γ of type A or $\check{A}$, and any $S_{p,q}$. If γ is of type A then Q is the double point of the center of γ, while if γ is of type $\check{A}$, then v is the double line of the center of γ.

13.3. Remark

The reason for taking 1/2 in the symbolic representation of L and $\check{L}$ as conditions of type $S_{0,1/2}$ and $S_{1/2,0}$ will be seen in next section. Here let us only say that the global characteristic numbers of $S_{p,q}$ are (2q,2p), and that this is again also true for these special cases.

§14. Conics satisfying five independent conditions

In this section we will consider the problem of determining the number of conics properly satisfying five conditions under the assumption that their data are in independent general position. In other words, the problem we want to solve is the determination of the number of conics properly satisfying $\sigma_i K_i$, $1 \leq i \leq 5$, where K_i are conditions and where $\sigma = (\sigma_1, \ldots, \sigma_5) \in G^5$ varies in a suitable open set of G^5 (cf. Remark 7.3).

14.1. Proposition

Given degeneration free reduced conditions K_i, $1 \leq i \leq 5$, there exists a non-empty open set U in G^5 such that for any $\sigma = (\sigma_1, \ldots, \sigma_5) \in U$ the conics properly satisfying $\sigma_i K_i$, $1 \leq i \leq 5$, (a) are non-degenerate and isolated components of multiplicity one of the intersection $\sigma_1 K_1 \cap \ldots \cap \sigma_5 K_5$, hence finite in number, and (b) this number is independent of $\sigma \in U$.

Proof.

Applying Kleiman's theorem (quoted as theorem 7.1) successively on $\mathbf{W}_o = \mathbf{W}-(\mathbf{A} \cup \check{\mathbf{A}})$, $\mathbf{A}-\mathbf{B}$, $\check{\mathbf{A}}-\mathbf{B}$, and $\mathbf{B}$ one obtains that there exists a non-empty open set $U \subseteq G^5$ such that if $\sigma = (\sigma_1, \ldots, \sigma_5) \in U$ and

$$I_\sigma := \sigma_1(K_1) \cap \ldots \cap \sigma_5(K_5)$$

then

(i) $I_\sigma \cap \mathbf{W}_o$ is a finite set and each point of this set has multiplicity one in the intersection,

(ii) $I_\sigma \cap ((\mathbf{A}-\mathbf{B}) \cup (\check{\mathbf{A}}-\mathbf{B})) = \emptyset$, and

(iii) $I_\sigma \cap \mathbf{B}$ is either empty or at least one of the conditions contains $\mathbf{B}$.

So we see that for $\sigma \in U$

$$I_\sigma = (I_\sigma \cap \mathbf{W}_o) \cup (I_\sigma \cap \mathbf{B}).$$

Now the elements of $I_\sigma \cap \mathbf{W}_o$ are non-degenerate and hence properly satisfy the five conditions, whereas $I_\sigma \cap B$ is either empty or consists of improper solutions for at least one condition. This proves (a).

In order to prove (b), set $K_i^o = K_i \cap W_o$ and define

$$Z \subset K_1^o \times \ldots \times K_5^o \times G^5$$

as the closed subvariety defined by the relations

$$\sigma_1(x_1) = \ldots = \sigma_5(x_5) .$$

Let $p\colon Z \longrightarrow G^5$ be the restriction to Z of pr_{G^5}. Given $\sigma \in G^5$, the closed set $p^{-1}(\sigma)$ can be identified with $I_\sigma \cap W_o$. Since for $\sigma \in U$ this set is finite it follows (for instance using [S], Ch. II, §5) that the cardinal of this set is constant on some non-empty open set contained in U. This completes the proof.

14.2. Remark

Proposition 14.1 can be extended at once to non-reduced degeneration free conditions except for the multiplicity one property of part (a). □

14.3. Corollary

Given five reduced conditions $K_1,\ldots,K_5$ there exists a non-empty open set $U \subseteq G^5$ such that for any $\sigma = (\sigma_1,\ldots,\sigma_5) \in U$ the clousure of the set of non-degenerate conics of $\sigma_1(K_1) \cap \ldots \cap \sigma_4(K_4)$ is a (1-dimensional) system of conics Γ and, moreover, the set of conics which properly satisfy $\sigma_1(K_1),\ldots,\sigma_5(K_5)$ is the set of conics in Γ which properly satisfy $\sigma_5(K_5)$.

Proof

Let $f\colon G^5 = G^4 \times G \longrightarrow G^4$ be the projection onto the first factor. For simplicity we will write, for $\sigma \in G^5$, $\sigma = (\sigma', \sigma_5)$, so that $\sigma' = f(\sigma)$. Let U' be a non-empty open set of G^5 satisfying 14.1. Since $f|_{U'}$ is dominant, using once more Kleiman's theorem there exists a non-empty open set $V \subseteq f(U')$ such that for any $\sigma' = (\sigma_1,\ldots,\sigma_4) \in V$

$$\Gamma^{o}_{\sigma'} := \sigma_1(K_1) \cap \ldots \cap \sigma_4(K_4) \cap \mathbf{W}_o$$

is a reduced curve and hence the clousure of $\Gamma^{o}_{\sigma'}$ in $\mathbf{W}$, which we will denote by $\Gamma_{\sigma'}$, is a system of conics. Now take $U = U' \cap f^{-1}(V)$. We claim that given $\sigma' \in V$ there exists $\sigma_5 \in G$ such that $(\sigma',\sigma_5) \in U$ and the number of conics properly satisfying $\sigma_i(K_i)$, $i = 1,\ldots,5$, agrees with the number of conics in Γ which properly satisfy $\sigma_5(K_5)$. This claim follows from the following two observations.

The first is that if a conic properly satisfies $\sigma_i(K_i)$, $1 \leqslant i \leqslant 5$, for $\sigma = (\sigma_1,\ldots,\sigma_5) \in U$, then this conic obviously is on $\Gamma_{\sigma'}$ and properly satisfies $\sigma_5(K_5)$. The second is that given $\sigma' = (\sigma_1,\ldots,\sigma_4) \in V$ there exists a non-empty open set U'' of G such that for any $\sigma_5 \in U''$ the conics in $\Gamma_{\sigma'}$ properly satisfying $\sigma_5(K_5)$ are non-degenerate (see 9.3.1), and consequently these conics properly satisfy $\sigma_1(K_1),\ldots,\sigma_4(K_4)$ as well. The existence of the σ_5 in the claim follows because $U \cap (\sigma' \times U'')$ is non-empty.

Now by proposition 14.1, and for $\sigma \in U$, the number n_σ of conics properly satisfying $\sigma_i(K_i)$, $i=1,\ldots,5$, is independent of σ. On the other hand, given $\sigma' \in V$, the number of conics in $\Gamma_{\sigma'}$ properly satisfying $\sigma_5(K_5)$ is independent of σ_5, by definition of proper intersection multiplicity (see 9.1). Let $n_{\sigma'}$ be this number. Then by the claim before we have that $n_\sigma = n_{\sigma'}$ for any $\sigma \in U$.

In order to end the proof, given $\sigma \in U$ let S_σ be the set of conics properly satisfying $\sigma_i(K_i)$, $i=1,\ldots,5$, and S'_σ be the set of conics in $\Gamma_{\sigma'}$ properly satisfying $\sigma_5(K_5)$. Since $S_\sigma \subseteq S'_{\sigma}$, because $S_\sigma \subseteq \mathbf{W}_o$ and $S_\sigma \cap \mathbf{W}_o = S'_\sigma \cap \mathbf{W}_o$, the equality $S_\sigma = S'_\sigma$ follows because $n_\sigma = n_{\sigma'}$. □

Thus we observe that in the problem of determining the number of conics properly satisfying five conditions, the result is a 5-linear function of the strict equivalence classes of the conditions. Therefore, by 12.3, the problem is reduced to the computation of the number of conics properly satisfying five elementary conditions.

In order to solve this problem it is convenient to prove first two auxiliary results. We recall that the elementary conditions are the $S_{p,q}$, (p,q) a pair of coprime positive integers, and also L and $\check{L}$, which are denoted by $S_{0,1/2}$ and $S_{1/2,0}$, respectively.

14.4. Lemma

Let S_{p_i,q_i}, $i=1,\dots,4$, be four elementary conditions ordered so that $p_1/q_1 \leqslant p_2/q_2 \leqslant p_3/q_3 \leqslant p_4/q_4$. Then there exists a non-empty open set V in G^4 such that for any $\sigma' = (\sigma_1,\dots,\sigma_4) \in V$ the clousure of the set of non-degenerate conics satisfying the four conditions $\sigma_i(S_{p_i,q_i})$ is a system of conics $\Gamma_{\sigma'}$ in **W** such that the ratio of order to class of any of its degenerations is either p_2/q_2 or p_3/q_3.

Proof

That there exists a non-empty open set V' in G^4 such that the closure of the set of non-degenerate conics satisfying the four conditions $\sigma_i(S_{p_i,q_i})$ is a system of conics $\Gamma_{\sigma'}$ is once again a direct consequence of Kleiman's theorem.

Now given $\sigma' \in V'$, let γ be a degeneration of $\Gamma_{\sigma'}$ of order m and class n. The set of $\sigma' \in V'$ for which $m/n < p_2/q_2$ is a proper closed subset of V', because in such a case, by proposition 13.1, the double focus of the center of γ would belong to the three lines with respect to which the conditions $\sigma_i(S_{p_i,q_i})$, $i=2,3,4$, are defined. By duality, the set of $\sigma' \in V'$ such that $p_3/q_3 < m/n$ is also a proper closed subset of V'. Finally, the set of $\sigma' \in V'$ such that $p_2/q_2 < m/n < p_3/q_3$ is a proper closed subset of V', because, by proposition 13.1 again, in this case the double line of the center of γ would pass through the two points with respect to which $\sigma_1(S_{p_1,q_1})$ and $\sigma_2(S_{p_2,q_2})$ are defined, so that the first two lines and the line joining the last two points are concurrent. □

14.5. Lemma

In the same situation as in lemma 14.4, let μ, ν be the global characteristic numbers of $\Gamma_{\sigma'}$, $\sigma' \in V'$. Denote by ap_2, aq_2 the sums of orders and classes, respectively, of the degenerations of $\Gamma_{\sigma'}$ for which the ratio of order to class is p_2/q_2. Likewise, let $a'p_3$, $a'q_3$ be the analogous sums for the degenerations whose ratio of order to class is p_3/q_3. Then

$$ap_2 + a'p_3 = 2\mu - \nu$$
$$aq_2 + a'q_3 = 2\nu - \mu.$$

Proof

By lemma 14.4 the sum M of the orders of all degenrations of Γ is equal to $ap_2+a'p_3$, and the sum N of the corresponding classes is $aq_2+a'q_3$. But by 5.6 we know that $M = 2\mu-\nu$ and $N = 2\nu-\mu$. □

Notice that 14.5 really allows to determine the sum of orders and the sum of classes of the degenerations of each type. This is clear if $p_2/q_2 \neq p_3/q_3$. But if $p_2/q_2 = p_3/q_3$, there is only one type of degeneration, so it suffices to determine $a+a'$, which can be done using either equation.

14.6. Theorem (Halphen's second formula)

Given five elementary conditions S_{p_i,q_i}, $1 \leqslant i \leqslant 5$, ordered in such a way that $p_i/q_i \leqslant p_{i+1}/q_{i+1}$ for $1 \leqslant i \leqslant 4$, there exists a non-empty open set U in G^5 such that for any $\sigma = (\sigma_1,\ldots,\sigma_5) \in U$ the number N of conics properly satisfying the conditions $\sigma_i(S_{p_i,q_i})$ is finite and given by the formula

$$N = 8(p_1+2q_1)(p_2+2q_2)(p_3+q_3)(2p_4+q_4)(2p_5+q_5).$$

Proof

Let $K_i = S_{p_i,q_i}$. Then we can determine a non-empty open set U' in G^5 such that for any five conditions chosen among the K_i, $1 \leq i \leq 5$, L and $\check{L}$, proposition 14.1 and corollary 14.3 apply. In particular for $\sigma \in U'$ there are only finitely many conics properly satisfying $\sigma_i(K_i)$, $1 \leq i \leq 5$, and the number N of such conics does not depend on $\sigma \in U'$. Now in order to compute N we will proceed recursively on the number of conditions L or $\check{L}$ that appear among the five elementary conditions K_i. If each of the five conditions is of type L or $\check{L}$, then there are no improper solutions and the number N can take on the following values:

$$L^5 = 1, \quad L^4\check{L} = 2, \quad L^3\check{L}^2 = 4, \quad L^2\check{L}^3 = 2, \quad L\check{L}^4 = 2, \quad \check{L}^5 = 1.$$

As it is checked immediately these values agree with the answer given by the claimed formula.

Therefore we may assume that among the conditions $K_i = S_{p_i,q_i}$ there is at least one, say K_5, which is neither of type L nor of type $\check{L}$ (so we do not assume so far that the ratios p_i/q_i are non-decreasing with respect to i). Given $\sigma \in U'$, let $\Gamma_{\sigma'}$ be the system of conics determined by the conditions $\sigma_i(K_i)$, $i < 5$, where $\sigma' \in G^4$ is the result of dropping σ_5 in σ. Now by recursion the claimed formula holds true for the conditions K_i, $i < 5$, together with either $S_{0,1/2}$ or $S_{1/2,0}$, and for some non-empty open set U that we can assume is contained in U'. Without loss of generality we may assume that $p_1/q_1 \leq p_2/q_2 \leq p_3/q_3 \leq p_4/q_4$ and so for $\sigma \in U$ the number μ of conics properly satisfying $\sigma_i(K_i)$, $i < 5$, and $\sigma_5(S_{0,1/2})$ will be

$$\mu = 8(p_1+2q_1)(p_2+q_2)(2p_3+q_3)(2p_4+q_4)$$

and the number ν of conics properly satisfying $\sigma_i(K_i)$, $i < 5$, and $\sigma_5(S_{1/2,0})$ will be

$$\gamma = 8(p_1+2q_1)(p_2+2q_2)(p_3+q_3)(2p_4+q_4) .$$

But, by the last part of 14.3, μ and ν are the number of conics in $\Gamma_{\sigma'}$ properly satisfying $\sigma_5(S_{0,1/2})$ and $\sigma_5(S_{1/2,0})$, respectively, and so (μ,ν) is the pair of global characteristic numbers of $\Gamma_{\sigma'}$. Using 14.5 a straightforward computation shows that if

$$a = 8(p_1+2q_1)(2p_3+q_3)(2p_4+q_4)$$

and

$$a' = 8(p_1+2q_1)(p_2+2q_2)(2p_3+q_3)$$

then ap_2, aq_2 are the sums of the orders and classes, respectively, of the degenerations of $\Gamma_{\sigma'}$ for which the ratio of order to class is p_2/q_2 and $a'p_3$, $a'q_3$ are to analogous sums for p_3/q_3. In case $p_2/q_2 = p_3/q_3$, then $(a+a')p_2$, $(a+a')q_2$ are the sums of orders and classes, respectively, of all degenerations.

Now given $\sigma \in U$, and according to 14.3, the conics that properly satisfy the five conditions $\sigma_i(K_i)$ are the conics in $\Gamma_{\sigma'}$ that properly satisfy $\sigma_5(K_5)$, and so we may compute the number N of such conics using Halphen's first formula 9.2. In principle we should consider five cases, according to the position of p_5/q_5 with respect to the ratios p_i/q_i, $i<5$. However, by duality the number N must be unaffected by interchanging the p's and q's, and so we only need consider the following two cases.

(a) $p_5/q_5 \geqslant p_3/q_3$.

In this case the ratio m/n of order to class for any degeneration of $\Gamma_{\sigma'}$, which equals p_2/q_2 or p_3/q_3, will be not greater than p_5/q_5, so that $\min(mq_5,np_5) = mq_5$. By Halphen's first formula 9.2 the number of proper solutions is given by

$$2q_5\mu + 2p_5\nu - ap_2q_5 - a'p_3q_5 = (2p_5+q_5)\nu .$$

Substituting ν for the expression found above we get the claimed formula.

(b) $p_2/q_2 \leqslant p_5/q_5 \leqslant p_3/q_3$.

In this case for the degenerations of Γ such that the ratio m/n equals p_2/q_2 we get $\min(mq_5, np_5) = mq_5$, while for the degenerations such that $m/n = p_3/q_3$ we get $\min(mq_5, np_5) = np_5$. Thus the number of solutions will be

$$2q_5\mu + 2p_5\nu - ap_2q_5 - a'q_3p_5 .$$

Substituting μ, ν, a and a' by the expresions found before we come up again with the claimed formula, up to reindexing. This completes the proof. □

14.7. Corollary (of the proof)

Given four conditions S_{p_i,q_i}, $1 \leqslant i \leqslant 4$, ordered in such a way that p_i/q_i is non-decreasing with i, assume that the data used to define these conditions has been taken generically and independently and let Γ be the system of conics they determine (cf. 14.3). Then

(a) The global characteristic numbers (μ, ν) of Γ are given by the formulae

$$\mu = 8(p_1+2q_1)(p_2+q_2)(2p_3+q_3)(2p_4+q_4)$$
$$\nu = 8(p_1+2q_1)(p_2+2q_2)(p_3+q_3)(2p_4+q_4)$$

(b) The ratio of order to class for any degeneration of Γ can only be p_2/q_2 or p_3/q_3. If $p_2/q_2 < p_3/q_3$ then the sum of the orders and the sum of classes for the degenerations whose ratio of order to class is p_2/q_2 are, respectively, ap_2 and aq_2, where

$$a = 8(p_1+2q_1)(2p_3+q_3)(2p_4+q_4) ;$$

the analogous sums for the degenerations whose ratio of order to class is p_3/q_3 are $a'p_3$ and $a'q_3$, respectively, where

$$a' = 8(p_1+2q_1)(p_2+2q_2)(2p_4+q_4) .$$

If $p_2/q_2 = p_3/q_3$ then the sum of orders and the sum of classes for all degenerations are $(a+a')p_2$ and $(a+a')q_2$, where a and a' have the same expression as above. □

We end this section with an example.

14.8. Example (Halphen)

Consider five conditions as described in example (K_2) of 11.3 defined with respect to five conics in general position. Halphen's second formula gives 1296 as the number of proper solutions, whereas Chasles' formula gives 3264. This last number, however, does not have enumerative significance because all degenerate conics of type B are (improper) solutions.

It may be appropriate to recall that 3264 is the number of conics that are tangent to five conics in general position, which of course is also the answer given by Halphen's second formula (actually Chasles' theory already gives this answer (only L's and $\check{L}$'s are involved)) and that one of the motivations for Chasles' theory was to improve Steiner and De Jonquières' approach which gave $6^5 = 7776$ as the number of conics tangent to five given conics in general position.

§15. Cycles on $\mathbf{W}$ and on $\mathbf{B}$

In this section we define a number of cycles that are relevant for our purposes, study their classes in the corresponding Chow ring, and establish a number of relations among them. Most of the computations are straightforward and are left to the reader. In a few cases we sketch a proof.

Given a point P in $\mathbb{P}_2$ we have denoted the cycle on $\mathbf{W}$ of conics that go through P by L_P. The class of this cycle in $A^1(\mathbf{W})$ is denoted by L. Given two points P,Q the cycle on $\mathbf{W}$ of conics that harmonically divide PQ also represents L and is denoted by $L_{P,Q}$.

Dually, given a line u, $\check{L}_u$ denotes the cycle on **W** of conics tangent to u, and $\check{L}$ is the class of $\check{L}_u$. This class is also represented by $\check{L}_{u,v}$, conics whose tangents from the intersection point of u and v harmonically divide the pair u,v.

We know that $A \sim 2\check{L}-L$, $\check{A} \sim 2L-\check{L}$, and that in fact L, $\check{L}$ is a free **Z**-basis of $A^1(\mathbf{W})$. Given a divisor D on **W**, if $K \sim aL+\check{a}\check{L}$, we will denote this cycle by $(a,\check{a})$ and write $K \sim (a,\check{a})$.

Since **W** is the blowing up of $\mathbb{P}_5$ along the Veronese surface V (see §2), from ([B], Ch.0; cf. also [F], 6.7 and Ex. 8.3.9) it follows that the Chow ring $A^{\cdot}(\mathbf{W})$ of **W** is isomorphic to the even cohomology ring $H^{2\cdot}(\mathbf{W})$ and that L^2, $\frac{L\check{L}}{2}$, $\check{L}^2$ is a free **Z**-basis of $A^2(\mathbf{W})$. Poincaré duality on cohomology tells us that $A^i(\mathbf{W})$ and $A^{5-i}(\mathbf{W})$ are dual under the intersection number product. We will denote by $S,T,\check{S}$ the basis of $A^3(\mathbf{W})$ that is the dual basis of L^2, $\frac{L\check{L}}{2}$, $\check{L}^2$ and by Γ and $\check{\Gamma}$ the basis of $A^4(\mathbf{W})$ that is the dual basis of L, $\check{L}$.

If K is a codimension 2 cycle and $K \sim aL^2 + b(\frac{L\check{L}}{2}) + \check{a}\check{L}^2$, then we will denote K by $(a,b,\check{a})$ and will also write $K \sim (a,b,\check{a})$. For codimension 3 cycles K, we will write $K \sim (a,b,\check{a})^T$ if $K \sim aS+bT+\check{a}\check{S}$. A similar notation will be used for codimension 4 cycles, so that $K \sim (a,\check{a})^T$ means that $K \sim a\Gamma+\check{a}\check{\Gamma}$. In this way the intersection number of $(a,b,\check{a})$ and $(c,d,\check{c})^T$ is $ac+bd+\check{a}\check{c}$.

On **B** we define the cycle ℓ_P of double lines with double focus that pass through P and will set ℓ to denote its class in $A^1(B)$. Dually, $\check{\ell}_u$ is the cycle on **B** of double lines with double focus lying on u, and $\check{\ell}$ its class in $A^1(B)$.

From the fact that **B** is a projective line bundle over the Veronese surface V it follows that the Chow ring of **B** is isomorphic to the even cohomology of B and also that the classes ℓ and $\check{\ell}$ form a free $\mathbb{Z}$-basis of $A^1(B)$, while $\check{\ell}^2$ and ℓ^2 form the basis of $A^2(\mathbf{W})$ that is dual of ℓ and $\check{\ell}$ by next relations.

15.1. (1) $\ell^3 = \check{\ell}^3 = 0$

(2) $\ell^2\check{\ell} = \ell\check{\ell}^2 = 1.$ □

15.2. Let $j\colon \mathbf{B} \longrightarrow \mathbf{W}$ be the inclusion. Then

(1) $j^*L = 2\ell$

(2) $j^*\check{L} = 2\check{\ell}$

Proof (cf. also 15.3.(2))

Consider the commuting diagram

$$\begin{array}{ccc} \mathbf{B} & \xrightarrow{\;j\;} & \mathbf{W} \\ {\scriptstyle \bar{p}}\downarrow & & \downarrow{\scriptstyle p} \\ V & \xrightarrow{\;i\;} & \mathbb{P}_5 \end{array}$$

where $i\colon V \longrightarrow \mathbb{P}^5$ is the inclusion of V in $\mathbb{P}_5$ and $\bar{p} = p|_{\mathbf{B}}$. Then if H is a hyperplane in $\mathbb{P}_5$,

$$j^*L = j^*p^*H = \bar{p}^*(2\bar{\ell}) \qquad (\bar{\ell} \text{ the class of a line in } \mathbb{P}_2 \xrightarrow[\sim]{v} V)$$
$$= 2\ell. \quad \square$$

The rest of the section is a list of useful cycles and relations among them.

Codimension 1 cycles

$A \sim$	$(-1,2)$	pairs of lines
$\check{A} \sim$	$(2,-1)$	double lines

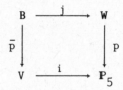

$S_{p,q} \sim (2q,2p)$ (defined in 11.1)

$\sim q\check{A} + (2p+q)\check{L} \sim pA + (p+2q)L.$

Codimension 2 cycles

$A_P = A \cdot L_P$	$\sim (-1,4,0)$	pairs of lines, one through P
$\check{A}_u = \check{A} \cdot \check{L}_u$	$\sim (0,4,-1)$	pairs of points, one on u
A_u	$\sim (0,-1,1)$	pairs of lines with double point on u
$\check{A}_P$	$\sim (1,-1,0)$	double lines thorugh P
$j_* B$	$\sim (-2,10,-2)$	

15.3. (1) $\check{A}.L_{P,Q} = \check{A}_P + \check{A}_Q$

(2) $\check{A}.L_P = 2\check{A}_P$

(3) $A.\check{L}_u = 2A_u$

(4) $A.\check{L}_{u,v} = A_u + A_v$

(5) $A.S_{p,q} = q\,B + (4p+2q)A_u \sim (-2q, 8q-4p, 4p)$

(6) $\check{A}.S_{p,q} = p\,B + (2p+4q)\check{A}_P \sim (4q,\ 8p-4q,\ -2p)$. □

Codimension 3 cycles

15.4. (1) $L^3 \sim (1,1,4)^T$

(2) $L^2\check{L} \sim (2,2,4)^T$

(3) $L\check{L}^2 \sim (4,2,2)^T$

(4) $\check{L}^3 \sim (4,1,1)^T$. □

$B_P = j_*(\ell_P)$ — double lines with double focus, the line through P

$B_u = j_*(\check{\ell}_u)$ — double lines with double focus, the focus on line u

15.5. (1) $B_P = \frac{1}{2}(L_P.B) \sim (0,2,4)^T$, $\quad B_u = \frac{1}{2}(\check{L}_u.B) \sim (4,2,0)^T$

(2) $\check{A}_P.A = B_P$, $\quad A_P.\check{A} = B_P$

(3) $A_u.\check{A} = B_u$, $\quad \check{A}_u.A = B_u$

(4) $L_{P,Q}.B = B_P + B_Q$

(5) $L_{u,v}.B = B_u + B_v$. □

$F_P \sim (0,1,1)^T$	double lines with one of its foci at P
$\check{F}_u \sim (1,1,0)^T$	pairs of lines, one of them equal to u
$D_P \sim (1,0,0)^T$	pairs of lines with double point at P
$\check{D}_u \sim (0,0,1)^T$	double line u^2 with a pair of points

(These relations can be seen applying previous relations and Kleiman's theorem in A or $\check{A}$. For instance, that $D_P \sim (a,0,0)^T$ is obvious, and a is 1 because $L_Q.A = A_Q$, which by Kleiman's theorem in A implies that $D_P.L^2=1$).

The first of next series of relations is justified below, the others are either similar or easier.

$\Lambda_{u,P} \sim (2,1,0)^T$	pairs of lines, double point $O \in u$ harmonically dividing u and OP.
$\check{\Lambda}_{u,P} \sim (0,1,2)^T$	double lines v going through P, foci harmonically dividing P and u.v
$\Sigma_{P,u,v} \sim (1,1,1)^T$	clousure of the set of non-degenerate conics tangent to u at u.v and such that the polar line of P is v (here we assume $P \in u$, v general)
$T_{P,Q,R} \sim (1,1,1)^T$	clousure of the set of non-degenerate conics for which the triangle P,Q,R is self-polar.

(For the cycle $\Sigma_{P,u,v}$, see 19.8.(b); here we are going to sketch the proof of the first of these relations. Set-theoretically it is clear that $A_u \cap L_{P,Q} = D_Q \cup \Lambda_{u,P}$. On the other hand, $A_u.L_{P,Q} \sim (3,1,0)^T$ and $D_Q \sim (1,0,0)^T$, which implies that $\Lambda_{u,P}$ is either linearly equivalent to $(1,1,1)^T$ or to $(2,1,0)^T$. But $\Lambda_{u,P}.L^2$ is at least 2, as a direct geometric argument shows.

Therefore $\Lambda_{u,P} \sim (2,1,0)^T$ and the multiplicities of the two components of $A_u \cap L_{P,Q}$ are one each.)

Codimension 4 cycles

15.6 (1) $L^4 \sim (1,2)^T$

(2) $L^3\check{L} \sim (2,4)^T$

(3) $L^2\check{L}^2 \sim (4,4)^T$

(4) $L\check{L}^3 \sim (4,2)^T$

(5) $\check{L}^4 \sim (2,1)^T$. □

15.7 $\Gamma_P \sim j_*(\check{\ell}^2) \sim (2,0)^T$ — double lines with double focus at P

$\check{\Gamma}_u \sim j_*(\ell^2) \sim (0,2)^T$ — u^2 with a variable double focus on u

$\Gamma_{u,Q} \sim j_*(\ell\check{\ell}) \sim (2,2)^T$ — double lines through Q having double focus on u

(The first of these relations comes from the fact that $\check{\ell}^2$ is reduced, by Kleiman's theorem on B, and that set-theoretically coincides with Γ_P. Then $\Gamma_P.L = j_*(\check{\ell}^2).L = \check{\ell}^2.j^*L = 2\check{\ell}^2.\ell = 2.$)

15.8 $\Gamma_{P,u,v} \sim (1,0)^T$ — pairs of lines with double point at P harmonically dividing u and v (u,v lines through P)

$\check{\Gamma}_{u,P,Q} \sim (0,1)^T$ — double lines u^2, with 2 points on it harmonically dividing P and Q.

(the first of these is a pencil and the second a range).

§16. Multiplicity cycles and Noether's formula

The main goal of this section is to establish a generalized version of Noether's classical formula according to which the total intersection number of two curves on a non-singular surface is the sum of the product of their multiplicites at a given point and the total intersection number of their strict transforms under the blowing up of the surface at that point. This formula then allows to express the total intersection number of the two curves as the sum of the products of the multiplicities of the curves extended to all their intersection points, including the infinitely near ones.

For our generalization the main ingredient, which we describe next is the notion of multiplicity cycle of a given cycle along a codimension two non-singular subvariety.

16.1. Definition

Let W be a non-singular variety, $B \subset W$ a non-singular codimension two subvariety, and K an irreducible cycle on W such that $K \not\subset B$. Under these hypothesis we define the *multiplicity cycle* K_B of K at B as follows:

$$K_B = \sum_{i=1}^{s} \mu_i Z_i \quad ,$$

where $Z_1,\dots,Z_s$ are the irreducible excedentary components of the intersection scheme $K \cap B$, and where μ_i is the multiplicity of the ideal of $K \cap B$ in $\mathcal{O}_{Z_i,K}$. The class $[K_B]$ in $A^1(B)$ will be called *multiplicity class* of K at B.

16.2. Remarks

Since B has codimension two and $K \not\subset B$, the components of $K \cap B$ are either 2-codimensional in B (proper components) or 1-codimensional (excedentary components). By definition, $K_B = 0$ if B and K intersect properly.

On the other hand, the definition of multiplicity cycle can be extended by linearity to any pure dimensional cycle none of whose components is contained in B.

16.3. Example

If W is a smooth surface, B a point of W, and K a curve on W, then

$$K_B = \mu_B(K)\cdot B \quad ,$$

where $\mu_B(K)$ is the multiplicity of B on K.

The multiplicity cycle can also be described as follows. Let $\varepsilon:\tilde{W} \longrightarrow W$ be the blowing-up of W along B, $E = \varepsilon^{-1}(B)$ the exceptional divisor, $g = \varepsilon|_E\colon E \longrightarrow B$, and let $i\colon E \longrightarrow \tilde{W}$, $j\colon B \longrightarrow W$ be the inclusion maps. Then we have:

16.4. Theorem

If $\tilde{K}$ is the strict transform of K under ε then

$$K_B = g_* i^* \tilde{K} \quad .$$

Proof

Without loss of generality we may assume that K is irreducible. Then $\tilde{K}$ is the blowing-up of K along its subscheme $K\cap B$ and the exceptional divisor of this blowing-up is $K\cap E$ (cf. [Har], II, 7.15). Let $\eta\colon \tilde{K} \longrightarrow K$ be the restriction of ε to $\tilde{K}$. Then because E is a divisor on $\tilde{W}$, $[\tilde{K}\cap E] = i^*\tilde{K}$, so that $g_* i^*\tilde{K} = g_*[E\cap\tilde{K}] = \eta_*[E\cap\tilde{K}]$.

As before, let $Z_1,\dots,Z_s$ be the excedentary components of $K\cap B$. Let $\tilde{Z}_{ij}$ be the components of $\eta^{-1}(Z_i)$. Since Z_i is excedentary, all $\tilde{Z}_{ij}$ have the same dimension as Z_i, and since η is proper it follows ([F], 4.3.6) that if μ_i is the multiplicity of the ideal of $K\cap B$ in

$\mathcal{O}_{Z_i,K}$, and μ_{ij} the multiplicity of the ideal of $E \cap \tilde{K}$ in $\mathcal{O}_{\tilde{Z}_{ij},\tilde{K}}$, then

$$\mu_i = \sum_j \deg(\tilde{Z}_{ij}|Z_i) \ \mu_{ij} .$$

But since the ideal of $E \cap \tilde{K}$ in the domain $\mathcal{O}_{\tilde{Z}_{ij},\tilde{K}}$ is principal, μ_{ij} is in fact the coefficient of $\tilde{Z}_{ij}$ in $[E \cap \tilde{K}]$, so that

$$[E \cap \tilde{K}] = \sum_{i,j} \mu_{ij} Z_{ij} .$$

Then we have

$$\begin{aligned} g_* i^* \tilde{K} &= \eta_* [E \cap \tilde{K}] \\ &= \sum_{i,j} \mu_{ij} \eta_* [\tilde{Z}_{ij}] \\ &= \sum_{i,j} \mu_{ij} \deg(\tilde{Z}_{ij}/Z_i) \cdot Z_i \\ &= \sum \mu_i Z_i \\ &= K_B . \quad \square \end{aligned}$$

Now we turn to Noether's generalized formula. With the same notations as before, recall that ([B], Ch. 0).

$$A^r(\tilde{W}) \xrightarrow{\ \sim\ } A^r(W) \oplus A^{r-1}(B)$$

under the map $x' \longrightarrow (\varepsilon_* x', g_*(h.i^* x'))$, where $h \in A^1(E)$ is the class of the tautological line bundle $\mathcal{O}_E(1)$. The inverse isomorphism is given by $(x,y) \longmapsto \varepsilon^* x + i_* g^* y$. Furthermore, for any $e \in A^\cdot(E)$, $i^* i_*(e) = -h \cdot e$.

16.5. Proposition

$$\varepsilon^*[K] = [\tilde{K}] + i_* g^* [K_B]$$

Proof

Since

$$\varepsilon_*(\varepsilon^*[K] - [\tilde{K}]) = 0 ,$$

it follows that there exists a class $k_o \in A^{r-1}(B)$ such that

$$\varepsilon^*[K] - [\tilde{K}] = i_* g^*(k_o) .$$

Then, in $A^{r-1}(B)$,

$$\begin{aligned} k_o &= k_o \cdot 1_B \\ &= k_o \cdot g_* h \\ &= g_*(g^* k_o \cdot h) && \text{(projection formula)} \\ &= -g_*(i^* i_* g^* k_o) \\ &= g_* i^*([\tilde{K}] - \varepsilon^*[K]) \\ &= g_* i^*[\tilde{K}] \\ &= [K_B] \end{aligned}$$

because $g_*(i^* \varepsilon^* K) = g_*(g^* j^* K) = 0$.

16.6. Theorem (Noether's formula)

With the same notations as before, let K, K' be two irreducible cycles, not contained in B. Let r and r' be their codimensions and assume $r+r' = \dim W$. Let K_B and K'_B be the multiplicity cycles of K and K', respectively. Then if K and K' intersect properly

$$\int_W K \cdot K' = \int_B K_B \cdot K_B + \int_{\tilde{W}} [\tilde{K}] \cdot [\tilde{K}']$$

Proof

Let $k_o = [K_B]$, $k'_o = [K'_B]$. Then

$$\begin{aligned} \int_W K \cdot K' &= \int_{\tilde{W}} \varepsilon^*[K] \cdot \varepsilon^*[K'] \\ &= \int_{\tilde{W}} ([\tilde{K}] + i_* g^* k_o) \cdot ([\tilde{K}'] + i_* g^* k'_o) \\ &= \int_{\tilde{W}} [\tilde{K}] \cdot [\tilde{K}'] + \int_{\tilde{W}} [\tilde{K}] \cdot i_* g^* k'_o + \int_{\tilde{W}} [\tilde{K}'] \cdot i_* g^* k_o + \int_{\tilde{W}} i_* g^* k_o \cdot i_* g^* k'_o . \end{aligned}$$

On the other hand, since $\varepsilon_* i_* g^* k_o = j_* g_* g^* k_o = 0$,

$$0 = \int_W \varepsilon_* i_* g^* k_o \cdot [K']$$
$$= \int_W \varepsilon_* (i_* g^* k_o \cdot \varepsilon^* [K'])$$
$$= \int_{\tilde{W}} (i_* g^* k_o \cdot \varepsilon^* [K'])$$
$$= \int_{\tilde{W}} (i_* g^* k_o \cdot [\tilde{K}'] + i_* g^* k_o \cdot i_* g^* k'_o) ,$$

so that

$$\int_{\tilde{W}} i_* g^* k_o \cdot [\tilde{K}'] = -\int_{\tilde{W}} i_* g^* k_o \cdot i_* g^* k'_o .$$

Similarly,

$$\int_{\tilde{W}} i_* g^* k'_o \cdot [\tilde{K}] = -\int_{\tilde{W}} i_* g^* k_o \cdot i_* g^* k'_o .$$

Hence

$$\int_W K \cdot K' = \int_{\tilde{W}} [\tilde{K}] \cdot [\tilde{K}'] - \int_{\tilde{W}} i_* g^* k_o \cdot i_* g^* k'_o .$$

But

$$i_* g^* k_o \cdot i_* g^* k'_o = i_* (g^* k_o \cdot i^* i_* g^* k'_o)$$
$$= -i_* (g^* k_o \cdot g^* k'_o \cdot h)$$

and consequently

$$-\int_{\tilde{W}} i_* g^* k_o \cdot i_* g^* k'_o = \int_E g^* (k_o \cdot k'_o) \cdot h$$
$$= \int_B (k_o \cdot k'_o) \cdot g_* h$$
$$= \int_B k_o \cdot k'_o .$$

This completes the proof. □

16.7. Remark

If the pairs (K,K') and $(\tilde{K},\tilde{K}')$ intersect properly, then

$$\int_{\tilde{W}} [\tilde{K}]\cdot[\tilde{K}'] = \int_{\tilde{W}} \tilde{K}\cdot\tilde{K}' = \#_E\ \tilde{K}\cdot\tilde{K}' + \#_{\tilde{W}-E}\tilde{K}\cdot\tilde{K}'$$

$$= \#_E\ \tilde{K}\cdot\tilde{K}' + \#_{W-B}K\cdot K' \quad ,$$

and

$$\int_W K\cdot K' = \#_B\ K\cdot K' + \#_{W-B}\ K\cdot K' \ ,$$

so that in this case Noether's formula is equivalent to the relation

$$\#_B\ K\cdot K' = \int_B K_B\cdot K_B' + \#_E\ \tilde{K}\cdot\tilde{K}' \quad . \ \square$$

§17. Improper intersection numbers

Given a degeneration free condition K we will say that a conic C properly satisfies K iff K intersects properly the orbit of C. For codimension one conditions this definition agrees with the definition given in 6.1. (The philosophy behind this definition is that if K has codimension i, $1\leqslant i\leqslant 4$, and if K improperly intersects the orbit of C, then K behaves, as far as imposing "conditions" to C in enumerative problems goes, as a condition of codimension $< i$, so C cannot be counted to satisfy K qua condition of codimension i.)

If C improperly satisfies a condition K, then C is of type B, because if K is degeneration free it intersects properly $\mathbf{A}-\mathbf{B}$ and $\check{\mathbf{A}}-\mathbf{B}$.

Now given two conditions K and K' of codimension i and $i' = 5-i$, respectively, Kleiman's theorem, applied successively to $\mathbf{W}_o$, $\mathbf{A}-\mathbf{B}$ and $\check{\mathbf{A}}-\mathbf{B}$ allows us to conclude that there exists a non-empty open set U of G such that for $\sigma\in U$ the conics that properly satisfy K and $\sigma(K')$ are non-degenerate and finite in number, and if the conditions are reduced each counts with multiplicity one in the intersection of K and $\sigma(K')$. In this section we are interested in finding an explicit expression for this number. To do that we begin with a definition.

17.1. Definition

The *improper intersection number* $\mathrm{imp}(K,K')$ of K and K' is defined to be $\#_B K\cdot\sigma(K')$, where $\sigma \in G$ is generic. Then the difference

$$\int_W K.K' - \mathrm{imp}(K,K')$$

will be called the proper intersection number of K and K' and will be denoted by $p(K,K')$. From this definition and Kleiman's theorem one sees immediately that $p(K,K') = \#_{W_o} K.\sigma(K')$, for $\sigma \in G$ generic, and that if K and K' are reduced $p(K,K')$ is the number of distinct conics properly satisfying K and $\sigma(K')$.

In order to compute $\mathrm{imp}(K,K')$ we need to relate it to the intersection number of the multiplicity classes and some sort of intersection of their strict transforms under the blowing up of **W** along **B**. To this end we will first introduce a couple of auxiliary results.

Let W be a smooth variety and G an algebraic group acting on W. Let A_1 and A_2 be two G-invariant irreducible non-singular closed hypersurfaces of W.

17.2. Definition

We will say that A_1 and A_2 have *good crossing* if the following conditions hold:

(i) $B = A_1 \cap A_2$ is an orbit of G (hence smooth);

(ii) A_1 and A_2 meet transversally along B; and

(iii) For each $z \in B$, the group of linear automorphisms of $N_{B/W}(z)$ induced by the isotropy group G_z of z contains (and hence is equal to) the subgroup of $GL(N_{B/W}(z))$ of those automorphisms that leave invariant the 1-dimensional linear subspaces $N_{B/A_1}(z)$ and $N_{B/A_2}(z)$.

Now let $\varepsilon : \tilde{W} \longrightarrow W$ be the blowing up of W along B, E the exceptional divisor, $g: E \longrightarrow B$ the restriction of ε to E. For $i=1,2$, let $\tilde{A}_i$ be the proper transform of A_i. With these notations we have:

17.3. Lemma

$\tilde{A}_i$ and E have good crossing with respect to the natural extension to $\tilde{W}$ of the action of G on W. Moreover, if $B_i := \tilde{A}_i \cap E$, then $g|_{B_i} : B_i \longrightarrow B$ is a G-isomorphism $(i=1,2)$.

Proof

First we describe the action of G on $\tilde{W}$. Let $\sigma \in G$ and set

$$\rho(\sigma): \quad W \longrightarrow W$$

to denote the action of σ on W. Then

$$\varepsilon_\sigma : \rho(\sigma) \circ \varepsilon : \quad \tilde{W} \longrightarrow W$$

has the property that $\varepsilon_\sigma^{-1} \mathcal{J}_B \mathcal{O}_{\tilde{W}}$ is invertible, where $\mathcal{J}_B$ is the sheaf of ideals of B in W. Hence there exists a unique morphism $\bar{\rho}(\sigma): \tilde{W} \longrightarrow \tilde{W}$ such that

$$\rho(\sigma) \circ \varepsilon = \varepsilon \circ \bar{\rho}(\sigma)$$

(see [Har], Prop. 7.14, Ch. II). The uniqueness implies immediately that $\bar{\rho}: G \longrightarrow \mathrm{Aut}(\tilde{W})$ is a morphism of groups. Now we will see that the map $\tilde{\rho}: G \times \tilde{W} \longrightarrow \tilde{W}$ given by

$$\tilde{\rho}(\sigma, x) = \bar{\rho}(\sigma)(x)$$

is a morphism. Indeed, the map

$$\rho \circ (1 \times \varepsilon): G \times \tilde{W} \longrightarrow W,$$

where $\rho: G \times W \longrightarrow W$ is the map $(\sigma, x) \longrightarrow \sigma(x)$, lifts to a morphism

$$\rho': G \times \tilde{W} \longrightarrow \tilde{W}$$

again because of the universality of the blowing-up, and $\rho' = \tilde{\rho}$ as one sees restricting both maps to $\{\sigma\} \times \tilde{W}$ for any $\sigma \in G$.

Now the action of G on the exceptional divisor E is given by projectivizing the action of G on $N_{B/W}$. More explicitly, if $z \in B$, $\sigma \in G$,

$$d_z\rho(\sigma): T_zW \longrightarrow T_{\sigma(z)}W$$

induces a linear map

$$\delta_z(\sigma): N_{B/W}(z) \longrightarrow N_{B/W}(\sigma z)$$

and the restriction of $\tilde{\rho}(\sigma)$ to $E_z := g^{-1}(z)$ maps E_z to $E_{\sigma(z)}$ and this map coincides with the projectivization of $\delta_z(\sigma)$ if, as usual, one identifies E_z with $\mathbb{P}(N_{B/W}(z))$ (see for instance [B-S], § 12).

We know that E is invariant under G. The invariance of $\tilde{A}_i$ follows immediately from the invariance of A_i. Moreover, $\tilde{A}_i$ is the blowing-up of A_i along $B \subset A_i$ and the exceptional locus of this blowing-up is $E \cap \tilde{A}_i$ ([Har], II, 7.15). Since B is a smooth hypersurface on A_i, it follows that $\tilde{A}_i$ is smooth and that $g: B_i \longrightarrow B$ is an isomorphism. (This isomorphism follows again from the local analysis explained below.)

It remains to be seen that $\tilde{A}_i$ and E intersect transversally along B_i, and that condition (iii) in the definition of good crossing is satisfied for $\tilde{A}_i$ and E at any point $z' \in B_i$. These assertions will be proved working with local equations for E and $\tilde{A}_i$ in a neighbourhood of z'.

Without loss of generality we may assume that $i=1$. So pick $z' \in B_1$

and set $z = g(z')$. Then the isotropy group $G_{z'}$ of $z' \in \tilde{W}$ coincides with the isotropy group G_z of $z \in W$. In fact the inclusion $G_{z'} \subseteq G_z$ is clear because g (or ε) is a G-morphism, and conversely, the elements of G_z leave invariant E_z and B_1, so that they must leave invariant $E_z \cap B_1 = \{z'\}$.

Consider now the action of an element $\sigma \in G_z = G_{z'}$ on $\mathcal{O}' := \mathcal{O}_{\tilde{W},z'}$. Let $\mathcal{O} := \mathcal{O}_{W,z}$ and set m' and m to denote the maximal ideals of $\mathcal{O}'$ and $\mathcal{O}$ respectively. Let $u_i \in m$ be a local equation of A_i, so that u_1, u_2 are local equations of B. Then $u_i^* := g^* u_i \in m'$ and, as is well known from the local description of the blowing-up, u_2^* is a local equation (at z') of E and there exists an element $v_1 \in m'$ suth that

$$(*) \qquad v_1 u_2^* = u_1^*$$

which is a local equation of $\tilde{A}_1$ and such that v_1 and u_2^* are part of a system of parameters for $\tilde{W}$ at z'. In particular $\tilde{A}_1$ is transversal to E at z'.

Now by the assumption that A_i is invariant it follows, given $\sigma \in G_z$, that

$$\rho(\sigma)^* u_i = t_i u_i \pmod{m^2},$$

where $t_i \neq 0$. And from condition (iii) in the definition of good crossing it follows that when σ varies in G_z then t_1, t_2 take over independently all non-zero values. This is so because $d_z u_1, d_z u_2$ is a basis of $N^*_{B/W}(z)$ and with respect to this basis the matrix of $\rho(\sigma)^*$ is

$$\begin{pmatrix} t_1 & 0 \\ 0 & t_2 \end{pmatrix}.$$

But now we will have

$$\tilde{\rho}(\sigma)^* v_1 = t v_1 \pmod{m'^2}, \quad t \neq 0 \quad (\text{invariance of } \tilde{A}_1)$$

$$\tilde{\rho}(\sigma)^* u_1^* = t_i u_i^*, \qquad i=1,2,$$

which together with the relation (*) imply that the matrix of $\tilde{\rho}(\sigma)^*$ with respect to the basis $d_{z'}v_1$, $d_{z'}u_2^*$ of $N^*_{B_1/\tilde{W}}(z')$ is

$$\begin{pmatrix} t_2/t_1 & 0 \\ 0 & t_2 \end{pmatrix} .$$

And it is clear, when σ varies in G_z, that t_2/t_1, t_2 take over independently all non-zero values, from which the good crossing of $\tilde{A}_1$ and E follows. □

17.4. Definition

Let W be a smooth variety and G an algebraic group acting on W. We will say that the action of G on W is good if there exist G-invariant smooth hypersurfaces $A_1,\ldots,A_k$ of W with the following properties:

(a) For all $i\neq j$, A_i and A_j have good crossing or $A_i \cap A_j = \emptyset$. We will set $B_{ij} = A_i \cap A_j$, so that B_{ij} is a codimension two orbit of G, if non-empty.

(b) The orbits of G are the sets $W - \bigcup_i A_i$, $A_i - \bigcup_j B_{ij}$ for $i=1,\ldots,k$, and B_{ij} for all $i\neq j$ such that $B_{ij} \neq \emptyset$, so that in particular any two of these sets are disjoint.

Now we have the following corollary of lemma 17.3:

17.5. Corollary

Assume G has good action on W. Let $B = \bigcup B_{ij}$ and $\varepsilon : \tilde{W} \longrightarrow W$ be the blowing-up of W along B. Then the natural action of G on $\tilde{W}$ is a good action.

Proof

Let $\tilde{A}_i$ be the strict trahsform of A_i under ε and if $B_{ij} \neq \emptyset$ let $E_{ij} = \varepsilon^{-1}(B_{ij})$, so that the union E of such E_{ij} is the exceptional divisor of ε. Then the hypersurfaces $\tilde{A}_1,\ldots,\tilde{A}_k$, and the E_{ij} for $B_{ij} \neq \emptyset$ are smooth, G-invariant and satisfy conditions (a) and (b). In fact (a) is a direct consequence of 17.3 and (b) follows immediately from the fact that $E_{ij}-\tilde{A}_i-\tilde{A}_j$ is an orbit of G.

Now let K and K' be degeneration-free cycles of codimension r and $n-r$, where $n = \dim(W)$, and assume again that G has a good action on W. Let B be the union of codimension two orbits, $\varepsilon : \tilde{W} \longrightarrow W$ the blowing-up of B and $\tilde{B}$ the union of codimension two orbits on $\tilde{W}$. Then we have the following:

17.6. Theorem

For a generic $\sigma \in G$, $\sigma(K)$ and K' intersect properly on W, $\sigma(\tilde{K})$ and $\tilde{K}'$ intersect properly on $\tilde{W}$, and

$$\#_B\ \sigma(K)\cdot K' = \int_B k_o \cdot k_o' + \#_{\tilde{B}}\ \sigma(\tilde{K})\cdot\tilde{K}' ,$$

where k_o and k_o' are the multiplicity classes of K and K' at B, respectively.

Proof

The first two statements follow from Kleinan's theorem applied to the orbits of G on W and $\tilde{W}$, respectively. The last follows readily from remark 16.7.

§18. Generalization of Halphen's formula to conditions of higher order

The first thing to do, in order to generalize Halphen's formula to conditions of higher order, is to define local characteristic numbers for such conditions.

Since we need work not only on **W** but also on successive blowing-ups of it along codimension two subvarieties, we will start with a smooth variety W in which two smooth irreducible hypersurfaces A_1, A_2 that intersect transversally along a (codimension two) smooth subvariety B are given.

In such a setting, let K be an irreducible subvariety of W such that $K \not\subseteq A_1, A_2$. We proceed to define local characteristic numbers for K at B. To do this, let $Z_1, \ldots, Z_s$ be the excedentary components of the intersection $B \cap K$. Since B has codimension two in W, we will have $\mathrm{cod}_K Z_i = 1$ for $1 \leqslant i \leqslant s$. Let $\mathcal{O}_i := \mathcal{O}_{K,Z_i}$, the local ring of K at Z_i, so that $\mathcal{O}_i$ is a 1-dimensional local domain. Let q_i be the ideal of $\mathcal{O}_i$ corresponding to $B \cap K$ and set $\mu_i = e_{\mathcal{O}_i}(q_i)$, the multiplicity of q_i in $\mathcal{O}_i$. Then the cycle

$$K_B = \sum_{i=1}^{s} \mu_i Z_i$$

has been called multiplicity cycle of K at B (cf. § 16).

Let $\bar{\mathcal{O}}_i$ be the integral clousure of $\mathcal{O}_i$ in the field $\mathbb{C}(K)$ of rational functions of K. Then $\bar{\mathcal{O}}_i$ has finitely many maximal ideals m_{ij}, $1 \leqslant j \leqslant r_i$, and the local rings $R_{ij} = (\bar{\mathcal{O}}_i)_{m_{ij}}$ are the discrete valuation rings of $\mathbb{C}(K)$ that dominate $\mathcal{O}_i$. We will write $v_{ij}: \mathbb{C}(K)^* \longrightarrow \mathbf{Z}$ to denote the valuation function corresponding to R_{ij}. For each i, let $x_i, y_i \in \mathcal{O}_i$ be such that (x_i) and (y_i) are the ideals defined by A_1 and A_2 in $\mathcal{O}_i$; in other words, x_i and y_i are local equations (in K) for $A_1 \cap K$ and $A_2 \cap K$ at the generic point of Z_i. Since B is the (complete) intersection of A_1 and A_2 we have that $q_i = (x_i, y_i)$.

18.1. Definitions

Let p_{ij}, q_{ij} be the pairs of coprime positive integers such that

$$p_{ij}/q_{ij} = v_{ij}(y_i)/v_{ij}(x_i) .$$

Then the pairs (p_{ij},q_{ij}), $1 \leqslant i \leqslant s$, $1 \leqslant j \leqslant r_i$, will be called *local characteristic pairs* of K at B, relative to A_1 and A_2. The pair (p_{ij},q_{ij}) will be said to correspond to v_{ij}. For a given characteristic pair (p_{ij},q_{ij}) there exists (clearly) a unique positive integer n_{ij} such that

$$v_{ij}(x_i) = n_{ij}\, p_{ij}, \qquad v_{ij}(y_i) = n_{ij}\, q_{ij} .$$

Let d_{ij} be the degree of the residue field of R_{ij} over the residue field of $\mathcal{O}_i$. Then the cycle $d_{ij}n_{ij}Z_i$ will be called R_{ij}*- multiplicity cycle* (or v_{ij}-multiplicity cycle) of K at B, and the cycle $K_{p,q} := \sum d_{ij}n_{ij}Z_i$, where the sum is extended to all pairs i,j such that $p_{ij} = p$, $q_{ij} = q$, will be called (p,q)*-multiplicity cycle* of K at B. Naturally, we set $K_{pq} = 0$ if (p,q) does not appear as a characteristic pair of K at B.

18.2. Theorem

$$K_B = \sum \min(p,q)K_{pq} ,$$

where the sum is extended to all pairs (p,q) of coprime positive integers.

Proof

We will see that Z_i appears with the same multiplicity on both sides. On the left hand side this multiplicity is $e_{\mathcal{O}_i}(q_i)$, and by ([Z-S], vol II, proof of Theorem 22 for d=1, p.294) we have that for any general C-linear combination f of x_i and y_i

$$e_{\mathcal{O}_i}(q_i) = \text{length } (\mathcal{O}_i/f\mathcal{O}_i)$$

(cf. also [F], Example 4.3.5). This in turn is equal to $\sum_j d_{ij}v_{ij}(f)$, as follows by [N], Theorems 6 and 8 (cf. also [F], Example A.3.1). Since $v_{ij}(f) = n_{ij}\min(p_{ij},q_{ij})$, we have that

$$e_{\mathcal{O}_i}(q_i) = \sum_j d_{ij}n_{ij}\min(p_{ij},q_{ij}) .$$

The theorem is now a direct consequence of this equality and the definition of $K_{p,q}$. □

18.3. Remark

If K is a purely dimensional cycle of W such that none of its components is contained in A_i, for $i=1,2$, then, by linearity, we can extend to K the definition of K_B, of local characteristic pairs, and of the cycles $K_{p,q}$. With this extension Theorem 18.2 is still valid.

The second step toward a generalization of Halphen's formula is to understand the relationship between the local characteristic numbers of K at B and those of the strict transform $\tilde{K}$ of K on the blowing-up $\varepsilon : \tilde{W} \longrightarrow W$ of W along B at the subvarieties $B_i := E \cap \tilde{A}_i$, $\tilde{A}_i$ the strict transform of A_i.

18.4. Proposition

The set of local characteristic pairs (p_1,q_1) of $\tilde{K}$ at B_1 (relative to $\tilde{A}_1$ and E) is in one-to-one correspondence with the set of local characteristic pairs (p,q) of K at B (relative to A_1 and A_2) such that $q > p$.

The relationship between (p_1,q_1) and the corresponding pair (p,q) is given by $p_1 = p$, $q_1 = q-p$ and for all such (p,q) the relation

$$\varepsilon_* \tilde{K}_{p,q-p} = K_{p,q}$$

holds.

There is a similar statement for B_2: in this case a pair (p,q), $p > q$, corresponds to $(p-q,q)$ and $\varepsilon_* \tilde{K}_{p-q,q} = K_{p,q}$.

Proof

Since $\varepsilon(\tilde{K} \cap B_1) \subseteq K \cap B$ and $\varepsilon|_{B_1}: B_1 \xrightarrow{\sim} B$, an excedentary component $\tilde{Z}$ of $\tilde{K} \cap B_1$ is mapped by ε to an excedentary component Z of $K \cap B$. Let $\mathcal{O} := \mathcal{O}_{Z,K}$, $\tilde{\mathcal{O}} := \mathcal{O}_{\tilde{Z},\tilde{K}}$, so that $\varepsilon^*: \mathcal{O} \hookrightarrow \tilde{\mathcal{O}}$ is a local morphism. Since $\varepsilon^*: \mathbb{C}(K) \xrightarrow{\sim} \mathbb{C}(\tilde{K})$, we will identify these two fields and in this way we see that any valuation v of $\mathbb{C}(\tilde{K})$ centered at $\tilde{Z}$ is also a valuation of $\mathbb{C}(K)$ centered at Z. The characteristic pair (p,q) of v at Z is defined by the relations $v(\bar{x}) = nq$, $v(\bar{y}) = np$, $(p,q) = 1$, where $\bar{x}$ and $\bar{y}$ are the elements in $\mathcal{O}$ corresponding to generators x,y of the ideals of A_1 and A_2 in $\mathcal{O}_{Z,W}$. Now the ideals of $\tilde{A}_1$ and E in $\mathcal{O}_{\tilde{Z},\tilde{W}}$ are generated by x/y and y respectively, and hence the characteristic pair (p_1,q_1) of v at $\tilde{Z}$ is given by the relations $v(\bar{y}) = n_1 p_1$, $v(\bar{x}/\bar{y}) = n_1 q_1$, $(p_1,q_1) = 1$. It turns out that $n_1 = n$, $p = p_1$ and $q = p_1+q_1$. In particular $q > p$.

Conversely, if Z is an excedentary component of $K \cap B$ and v is a valuation of $\mathbb{C}(K)$ centered at $\mathcal{O} := \mathcal{O}_{Z,K}$ such that $q > p$, where (p,q) is the characteristic pair of v at Z, then Z,v comes from some $\tilde{Z},v$ (necessarily unique) the way explained in the first paragraph. In order to see this, let U be an affine open set of K such that $U \cap Z \neq \emptyset$, $I_U(A_1 \cap U) = (\bar{x})$, $I_U(A_2 \cap U) = (\bar{y})$, where $\bar{x},\bar{y} \in A := \mathbb{C}[U]$. Then $\tilde{K}$ contains an affine open set U' such that $U' \cap B_1 \neq \emptyset$ and with $\mathbb{C}[U'] = A[\bar{x}/\bar{y}]$. Let R_v be the valuation ring of v. Then since $\mathcal{O} < R_v$, $A \subseteq R_v$. The hypothesis $q > p$ implies that $\bar{x}/\bar{y} \in m_v$, the maximal ideal of R_v. Thus

$$\tilde{\mathfrak{p}} := m_v \cap A[\bar{x}/\bar{y}]$$

is a prime ideal of $\mathbb{C}[U']$ such that $(\bar{y},\bar{x}/\bar{y}) \subseteq \tilde{\mathfrak{p}}$. This implies that if $\tilde{Z}$ is the closed set of $\tilde{K}$ corresponding to $\tilde{\mathfrak{p}}$ then $\tilde{Z} \cap U' \subseteq U' \cap B_1$ and so

$\tilde{Z} \subseteq B_1$. Since clearly $\varepsilon(\tilde{Z}) = Z$, $\tilde{Z}$, v satisfies the claimed properties.

To end the proof it is enough to observe that the pair Z, v contributes to $\tilde{K}_{p,q-p}$ as $\tilde{d}n\tilde{Z}$, where $\tilde{d} = [k(R_v) : k(\tilde{\mathcal{O}})]$, while Z, v contributes as dnZ, where $d = [k(R_v) : k(\mathcal{O})]$. But since $\varepsilon : \tilde{Z} \xrightarrow{\sim} Z$, $\tilde{d} = d$ and so the first contribution is $dn\tilde{Z}$, which ε_* maps to dnZ. □

We are now ready to prove the generalization of Halphen's formula announced before. We use the notations explained after remark 18.3.

18.5. Theorem (Halphen's formula for higher codimensions)

Let K and K' be effective cycles of codimension i and $i' = n-i$, $n = \dim(W)$, such that no component of either of them is contained in A_1 or A_2. Then there exists a non-empty open set U of G such that $\sigma(K)$ and K' intersect properly on W for any $\sigma \in U$ and

$$\#_B \sigma(K) \cdot K' = \sum \min(pq', qp') \int_B K_{p,q} \cdot K'_{p',q'} ,$$

where the sum is extended over all local characteristic pairs (p,q) and (p',q') of K and K', respectively, at B.

Proof

We may assume that K and K' are irreducible (remark 18.3). Using Kleiman's theorem one sees easily that there exists a non-empty open set U of G such that $\sigma(K)$ and K' intersect properly on W, and that $\sigma(\tilde{K})$ and $\tilde{K}'$ intersect properly on $\tilde{W}$, for any $\sigma \in U$. Then, by 16.7,

$$\#_B \sigma(K) \cdot K' = \int_B K_B \cdot K'_B + \#_E \sigma(\tilde{K}) \cdot \tilde{K}' ,$$

where K_B and K'_B are the multiplicity cycles of K and K' at B. We may assume that for $\sigma \in U$ the cycles $\sigma(\tilde{K})$ and $\tilde{K}'$ have no intersection points on $E - B_1 - B_2$. For $E' := E - B_1 - B_2$ is an orbit of G (17.5) and $\sigma(\tilde{K}) \cap E'$, $\tilde{K}' \cap E'$ have codimension i and $n-i$ in E', so that they do not intersect if σ is generic in G.

Therefore we may write

$$(*) \qquad \#_B \sigma(K)\cdot K' = \int_B K_B \cdot K'_B + \#_{B_1} \sigma(\tilde{K})\cdot \tilde{K}' + \#_{B_2} \sigma(\tilde{K})\cdot \tilde{K}' \ .$$

We will prove the theorem using this relation and induction on the maximum value m of the characteristic numbers of K at B.

If $m=0$ then $K \cap B$ does not have excedentary components and both members of the claimed formula are zero. So suppose $m > 0$. In the relation $(*)$ above, let m_i, $i=1,2$, be the maximum of the characteristic numbers of $\tilde{K}$ at B_i. Then, by 18.4, $m_i < m$ and hence by induction we may assume that

$$\#_{B_i} \sigma(\tilde{K})\cdot \tilde{K}'$$

can be expressed by means of the formula to be established, for all σ in some non-empty open set, which we can assume is U (shrinking the former U if necessary). So for $\sigma \in U$ we have

$$\#_{B_1} \sigma(\tilde{K})\cdot \tilde{K}' = \sum \min(\bar{p}\bar{q}', \bar{q}\bar{p}') \int_{B_1} \tilde{K}_{\bar{p},\bar{q}} \, \tilde{K}_{\bar{p}',\bar{q}'} \ ,$$

where the sum is extended to all characteristic pairs $(\bar{p},\bar{q})$ and $(\bar{p}',\bar{q}')$ of $\tilde{K}$ and $\tilde{K}'$, respectively, at B_1. But by 18.4 these characteristic pairs are of the form $(p,q-p)$, $(p',q'-p')$, where (p,q), (p',q') are the characteristic pairs of K and K', respectively, at B with $q > p$, $q' > p'$. Therefore

$$\#_{B_1} \sigma(\tilde{K})\cdot \tilde{K}' = \sum_{\substack{q>p \\ q'>p'}} \min(p(q'-p'),(q-p)p') \int_{B_1} \tilde{K}_{p,q-p} \cdot \tilde{K}'_{p',q'-p'} =$$

$$= \sum_{\substack{q>p \\ q'>p'}} \min(pq',qp') \int_B K_{p,q} \cdot K'_{p',q'} - \sum_{\substack{q>p \\ q'>p'}} pp' \int_B K_{p,q} \cdot K'_{p',q'} \ ,$$

where the sums are extended to characteristic pairs (p,q), (p',q') of K and K', respectively, with the restrictions made explicit below the summation signs.

Similarly,

$$\#_{B_2}\sigma(\tilde{K})\cdot\tilde{K}' = \sum_{\substack{q<p\\ q'<p'}} \min(pq',qp') \int_B K_{p,q}\cdot K'_{p',q'} - \sum_{\substack{q<p\\ q'<p'}} qq' \int_B K_{p,q}\cdot K'_{p',q'} .$$

On the other hand

$$\int_B K_B\cdot K_{B'} = \sum_{\substack{p,q\\ p',q'}} \min(p,q)\cdot\min(p',q') \int_B K_{p,q}\cdot K'_{p',q'} ,$$

by 18.2.

The claimed formula follows now easily by substituting the last three equalities in (*) and observing that $\min(p,q)\min(p',q')$ is equal to pp' if $q > p$ and $q' > p'$; to qq' if $q < p$ and $q' < p'$; and to $\min(pq',qp')$ in all other cases.

18.6. Remark

It is easy to see that when $i=1$, $i'=4$ and **W** is the variety of complete conics, then 18.5 gives Halphen's first formula. See also 20.2.

§ 19. Examples of higher order conditions

This section is devoted to the construction of certain cycles (in codimensions 2, 3 and 4) and to compute their Chow classes and multiplicity cycles. The leading idea is to find, for each codimension, cycles with the simplest (non-trivial) local characters.

Construction of the codimension 2 cycles $H_{p,q}$

Let P,Q be two distinct points of $\mathbb{P}_2$ and let u be a line going through P and not through Q. If p,q are coprime positive integers then we may use P and u to define the condition $S_{p,q}$ (cf. 11.1; as we explain there in it is not necessary to specify the additional two points on u and two lines through P required to define $S_{p,q}$). Let also denote by $L_{P,Q}$ the condition of harmonically dividing PQ. Then the intersection $L_{P,Q} \cap S_{p,q}$ has codimension two and contains $\check{A}_P$ (double lines through P) as a component.

19.1. Lemma

The multiplicity of $\check{A}_P$ in the intersection $L_{P,Q} \cap S_{p,q}$ is equal to q.

Proof

From the description of $S_{p,q}$ given at the end of 11.1 we know that it belongs to the pencil defined by the (linearly equivalent) divisors $q\check{A} + (2p+q)\check{L}_u$ and $pA + (2q+p)L_P$. Let us compute the multiplicity of $\check{A}_P$ in the intersection of these divisors with $L_{P,Q}$.

We have that

$$L_{P,Q}\cdot(q\check{A} + (2p+q)\check{L}_u) = q\, L_{P,Q}\cdot\check{A} + (2p+q)L_{P,Q}\cdot\check{L}_u$$

$$= q(\check{A}_P+\check{A}_Q) + (2p+q)L_{P,Q}\cdot\check{L}_u$$

(see §15 for the computation of the intersection products used here). Since $\check{L}_u$ does not contain $\check{A}_P$, it follows that $\check{A}_P$ has multiplicity q in $L_{P,Q}\cdot(q\check{A}+(2p+q)\check{L}_u)$.

Similarly,

$$L_{P,Q}\cdot(pA + (p+2q)L_P) = pL_{P,Q}\cdot A + (p+2q)L_{P,Q}\cdot L_P$$

and $\check{A}_P$ is not contained in $\mathbf{A}$, while $L_{P,Q} \cdot L_P = \check{A}_P + \frac{1}{2} L_P \check{L}_v$, where v is the line PQ. Thus $\check{A}_P$ has multiplicity p+2q in $L_{P,Q} \cdot (p\mathbf{A}+(p+2q)L_P)$. Consequently the multiplicity of $\check{A}_P$ in $L_{P,Q} \cdot S_{p,q}$ is q (the least of the two just computed intersection numbers). □

Thus $L_{P,Q} \cdot S_{p,q} - q\check{A}_P$ is an effective cycle that does not contain $\check{A}_P$ as a component. This cycle will be denoted by $H_{p,q}$.

19.2. Remark

$H_{p,q}$ is the clousure of the set of non-degenerate conics that harmonically divide PQ and satisfy $S_{p,q}$. Indeed, since $S_{p,q} \cap \mathbf{A} = \mathbf{B} \cup A_u$, $S_{p,q} \cap \check{\mathbf{A}} = \mathbf{B} \cup \check{A}_P$, $\mathbf{B} \notin L_{P,Q}$, and $A_u \notin L_{P,Q}$, we see that $\check{A}_P$ is the only degenerate component of $L_{P,Q} \cdot S_{p,q}$. □

19.3. Lemma

$H_{p,q} \cap \mathbf{B} = B_Q \cup \Gamma_{uQ} \cup \check{\Gamma}_u$, so that B_Q is the only excedentary component of $H_{p,q} \cap \mathbf{B}$.

Proof

First notice that set-theoretically we have

$$\begin{aligned}(H_{p,q}+q\check{A}_P) \cap \mathbf{A} &= S_{p,q} \cap L_{P,Q} \cap \mathbf{A} \\ &= (\mathbf{B} \cup A_u) \cap L_{P,Q} \\ &= B_P \cup B_Q \cup (A_u \cap L_{P,Q}) .\end{aligned}$$

Now let us compute the intersection multiplicity of $H_{p,q} + q\check{A}_P$ and $\mathbf{A}$ at B_P:

$$\begin{aligned}i_{B_P}((H_{p,q}+q\check{A}_P) \cdot \mathbf{A}) &= i_{B_P}(S_{p,q} \cdot (L_{P,Q} \cdot \mathbf{A})) \\ &= i_{B_P}((q\check{\mathbf{A}} + (2p+q)\check{L}_u) \cdot (L_{P,Q} \cdot \mathbf{A}))\end{aligned}$$

(because $S_{p,q}$ is in the pencil defined by $q\check{\mathbf{A}}+(2p+q)\check{L}_u$ and $p\mathbf{A}+(p+2q)L_P$

and the latter contains $\mathbf{A}$)

$$= q\ i_{B_P}(\check{\mathbf{A}}\cdot(L_{P,Q}\cdot\mathbf{A})) = q$$

(because $\check{\mathbf{A}}\cdot(L_{P,Q}\cdot\mathbf{A}) = L_{P,Q}\cdot\mathbf{B} = B_P + B_Q$ and $L_{P,Q}\cdot\check{L}_u\cdot\mathbf{A}$ does not contain B_P).

This implies that $H_{p,q} \not\supset B_P$. Indeed, otherwise $H_{p,q} \cap \mathbf{A}$ would contain B_P, which would force the multiplicity of B_P in $(H_{p,q}+q\check{A}_P) \cap \mathbf{A}$ to be greater than q (clearly $\check{A}_P \cap \mathbf{A}$ contains B_P). Therefore

$$H_{p,q} \cap \mathbf{A} = B_Q \cup (A_u \cap L_{P,Q})$$

and consequently

$$\begin{aligned} H_{p,q} \cap \mathbf{B} &= H_{p,q} \cap \mathbf{A} \cap \check{\mathbf{A}} \\ &= B_Q \cup (A_u \cap L_{P,Q} \cap \check{\mathbf{A}}) \\ &= B_Q \cup \Gamma_{uQ} \cup \check{\Gamma}_u\ . \quad \square \end{aligned}$$

19.4. Theorem

(a) $H_{p,q} \sim qL^2 + (4p+q)\frac{L\check{L}}{2}$

(b) For any pair of positive coprime integers p',q' we have that

$$(H_{p,q})_{p',q'} = \begin{cases} 0 & \text{if } (o',q') \neq (p,q) \\ B_Q & (\sim \ell \text{ in } A^1(\mathbf{B})) \text{ if } (p',q') = (p,q) \end{cases} .$$

Proof

(a) It is an immediate consequence of the definition of $H_{p,q}$.

(b) Since B_Q is the only excedentary component of $H_{p,q} \cap \mathbf{B}$, let us consider the local rings $\mathcal{O}$ and $\bar{\mathcal{O}}$ of B_Q in $\mathbf{W}$ and $H_{p,q}$, respectively. Let $f \in \mathcal{O}$ be a generator of the ideal defined by $L_{P,Q}$ and let $x,y \in \mathcal{O}$ be the generators of the ideals defined by $\mathbf{A}$ and $\check{\mathbf{A}}$, respectively, induced by the functions X,Y defined using u and P as in §4. Then $S_{p,q}$ defines an ideal in $\mathcal{O}$ generated by $\lambda x^p - y^q$, λ a non-zero scalar, and therefore the ideal of $H_{p,q}$ in $\mathcal{O}$ is $(f,\lambda x^p-y^q)$. It follows that $\bar{\mathcal{O}} = \mathcal{O}/(f,\lambda x^p-y^q)$. But $\mathcal{O}' = \mathcal{O}/(f)$ is a regular local ring of dimension 2 and $\bar{\mathcal{O}} = \mathcal{O}'/(\lambda x'^p-y'^q)$, where $x',y' \in \mathcal{O}'$ are the classes of $x,y \in \mathcal{O}$. In particular $\bar{\mathcal{O}}$ is a domain and so there is only one irreducible component K of $H_{p,q}$ containing B_Q. Thus $\bar{\mathcal{O}} = \mathcal{O}_{B_Q,K}$ and any local characteristic pair of $H_{p,q}$ will come from K. But now taking the completion of $\bar{\mathcal{O}}$ we see that there exists a unique valuation v of $\mathbb{C}(K)$ centered at $\bar{\mathcal{O}}$ and that such valuation satisfies $v(x) = q$, $v(y) = p$, $d=1$ (where d is the degree of the residue field of R_v over the residue field of $\mathcal{O}$). This completes the proof. □

The dual construction of $H_{p,q}$ leads to the construction of cycles that will be denoted $\check{H}_{q,p}$. The (dual) role of the point Q not on u is played by a line v not through P. With these notations we have:

19.4̌. Theorem

(a) $\check{H}_{p,q} \sim p\check{L}^2 + (p+4q)\frac{L\check{L}}{2}$

(b) For any pair of positive coprime integers (p',q')

$$(\check{H}_{p,q})_{p',q'} = \begin{cases} 0 & \text{si } (p',q') \neq (p,q) \\ B_v \ (\sim \check{\ell} \text{ in } A^1(B)) & \text{if } (p',q')=(p,q). \end{cases}$$ □

Construction of the codimension 3 cycles $G_{p,q}$

Fix three non-collinear points P,Q,Q' and let v,v',w be the lines PQ, PQ', and QQ', respectively.

19.5. Lemma

The cycle $\check{A}_P$ is a component of multiplicity one in the intersection of the cycles $L_{P,Q}$ and $L_{P,Q'}$.

Proof

It is easy to check that

$$L_{P,Q} \cap L_{P,Q'} \cap \mathbf{A} = B_P \cup D_P \cup \Lambda_{wP} \ .$$

Since

$$L_{P,Q} \cdot L_{P,Q'} \cdot \mathbf{A} \sim \begin{pmatrix} 3 \\ 3 \\ 4 \end{pmatrix}, \qquad B_P \sim \begin{pmatrix} 0 \\ 2 \\ 4 \end{pmatrix}, \quad D_P \sim \begin{pmatrix} 1 \\ 0 \\ 0 \end{pmatrix}, \quad \Lambda_{wP} \sim \begin{pmatrix} 2 \\ 1 \\ 0 \end{pmatrix},$$

actually we have

$$L_{P,Q} \cdot L_{P,Q'} \cdot \mathbf{A} = B_P + D_P + \Lambda_{wP} \ .$$

Now $\check{A}_P$ is clearly a component of $L_{P,Q} \cap L_{P,Q'}$. Let $m \geq 1$ be its multiplicity. Then since $\check{A}_P \cdot \mathbf{A} = B_P$, we see that the multiplicity of B_P in $L_{P,Q} \cdot L_{P,Q'} \cdot \mathbf{A}$ is at least m. Thus m=1. □

Set $G = L_{P,Q} \cdot L_{P,Q'} - \check{A}_P$. Then G is an effective cycle that does not contain $\check{A}_P$ as a component. Clearly $G \sim \frac{L\check{L}}{2}$. As a consequence of the previous lemma and its proof we have:

19.6. The intersection $G.\mathbf{A}$ is proper and $G.\mathbf{A} = D_P + \Lambda_{wP}$. □

19.7. Remark

The intersection $G.\check{\mathbf{A}}$ is also proper, as $L_{P,Q} \cdot L_{P,Q'}$ only has the component $\check{A}_P$ in $\check{\mathbf{A}}$. Thus we may say that G is determined by its intersection with the open set of non-degenerate conics. We will express this by saying that G is the clousure of the family of non-degenerate conics for which P,Q and P,Q' are conjugate pairs.

Now since on non-degenerate conics these conditions are equivalent to being conjugate the pairs of lines w,v and w,v', we have also the expression

$$G = \check{L}_{w,v} \cdot \check{L}_{w,v'} - A_w \quad .$$

Moreover,

$$G.\check{A} = \check{D}_w + \check{\Lambda}_{wP} \quad . \quad \square$$

19.8. Lemma

(a) $G.L_P \quad = D_P + \check{\Lambda}_{w,P}$

(b) $G.\check{L}_u \quad = D_P + \Sigma_{P,u,w}$

(c) $G.L_{Q,Q'} = \check{D}_w + T_{P,Q,Q'}$.

Proof

(a) By the definition of G, remark 19.7, and 19.6 we see that (set-theoretically)

$$G = L_{P,Q} \cap L_{P,Q'} \cap \check{L}_{w,v} \cap \check{L}_{w,v'}$$

and hence

$$G \cap L_P = D_P \cup \check{\Lambda}_{w,P} \quad .$$

But now

$$G \cdot L_P \sim \begin{pmatrix} 1 \\ 1 \\ 2 \end{pmatrix}, \qquad D_P \sim \begin{pmatrix} 1 \\ 0 \\ 0 \end{pmatrix}, \qquad \text{and} \qquad \check{\Lambda}_{w,P} \sim \begin{pmatrix} 0 \\ 1 \\ 2 \end{pmatrix}$$

and so (a) follows.

(b) It is clear that D_P is a component, and in fact the only degenerate component, of $G \cap \check{L}_u$. Let Σ be the clousure of the non-degenerate conics in $G \cap L_P$. It is easy to see that $\Sigma = \Sigma_{P,u,w}$. Now since

$$G.\check{L}_u \sim \begin{pmatrix} 2 \\ 1 \\ 1 \end{pmatrix}, \qquad D_P \sim \begin{pmatrix} 1 \\ 0 \\ 0 \end{pmatrix}, \qquad \text{and} \qquad \Sigma_{P,u,w} \sim \begin{pmatrix} \alpha \\ \beta \\ \gamma \end{pmatrix} ,$$

with $\alpha > 0$ (as it is easy to see), it turns out that actually $\alpha = \beta = \gamma = 1$, and that $G.L_u = D_P + \Sigma_{P,u,w}$.

(c) From $G + \check{A}_P = L_{P,Q} \cdot L_{P,Q'}$ it follows that

$$G.L_{Q,Q'} + \check{D}_V + \check{D}_{V'} = L_{P,Q}.L_{P,Q'}.L_{Q,Q'}$$

$$= T_{P,Q,Q'} + \check{D}_V + \check{D}_{V'} + \check{D}_W \quad ,$$

and this completes the proof. □

19.9. Lemma

The irreducible cycles D_P and $\check{\Lambda}_{w,P}$ are the degenerate components of the intersection $S_{p,q}.G$ and the corresponding multiplicities are $2p+q$ and q, respectively.

Proof

It is similar to the proof of 19.1 and we will omit it. In the computations (a) and (b) of lemma 19.8 are used. □

Thus the codimension 3 cycle

$$G_{p,q} := G.S_{p,q} - (2p+q)D_P - q\check{\Lambda}_{w,P}$$

is effective and has no degenerate component. The cycle $\check{G}_{q,p}$ is defined using the dual construction of $G_{p,q}$.

19.10. Theorem

(a) $G_{p,q} \sim (2p+q,\ 2p+q,\ 2p+2q)^T$.

(b) For any pair (p',q') of positive coprime integers,

$$(G_{p,q})_{p',q'} = \begin{cases} 0 & \text{if } (p',q') \neq (p,q) \\ \check{\Gamma}_w \ (\sim \ell^2 \text{ in } A^1(B)) & \text{if } (p',q') = (p,q). \end{cases}$$

(ǎ) $\check{G}_{p,q} \sim (2p+2q,\ p+2q,\ p+2q)^T$.

(b̌) For any pair (p',q') of positive coprime integers

$$(\check{G}_{p,q})_{p',q'} = \begin{cases} 0 \text{ if } (p',q') \neq (p,q) \\ \Gamma_P \ (\sim \check{\ell}^2 \text{ in } A^1(B)) \text{ if } (p',q') = (p,q). \end{cases}$$

Proof

Formula (a) can be obtained (recalling that $G \sim \frac{L\check{L}}{2}$, $S_{p,q} \sim 2qL+2p\check{L}$, $D_P \sim L^2$, and $\check{\Lambda}_{w,P} \sim 2L^2 + \frac{L\check{L}}{2}$) by a straightforward computation.

(b) Recall that G and $\mathbf{A}$ intersect properly and that $G.\mathbf{A} = D_P + \Lambda_{w,P}$ (see 19.6). Thus $\check{\mathbf{A}}$ properly intersects $G.\mathbf{A}$ and hence

$$G.\mathbf{B} = (G.\mathbf{A})\cdot\check{\mathbf{A}} = \Gamma_P + \check{\Gamma}_w + \Gamma_{w,P} ,$$

as $D_P.\mathbf{A} = \Gamma_P$, $\Lambda_{w,P}.\check{\mathbf{A}} = \check{\Gamma}_w + \Gamma_{w,P}$. In particular

$$G_{p,q} \cap \mathbf{B} \subseteq G \cap \mathbf{B} = \Gamma_P \cup \check{\Gamma}_w \cup \Gamma_{w,P} .$$

Now $\check{\Gamma}_w$ is a component of $G_{p,q} \cap \mathbf{B}$, and so an excedentary component. In fact it is enough to show that $\check{\Gamma}_w$ is contained in $G_{p,q}$, and this follows easily from the definition of G and $G_{p,q}$. The cycles Γ_P and $\Gamma_{w,P}$, however, are not contained in $G_{p,q}$ (lemma 19.11 below) and therefore $\check{\Gamma}_w$ is the only excedentary component of $G_{p,q} \cap \mathbf{B}$.

Since $\check{\Gamma}_w$ has multiplicity one in the intersection of G and $\mathbf{B}$, the local ring $\mathcal{O}_{\check{\Gamma}_w,G}$ is regular (of dimension 2). If we take equations X,Y of $\mathbf{B}$ at the generic point of $\check{\Gamma}_w$ as in §4, then the restrictions x, y of X,Y to G generate the maximal ideal of $\mathcal{O}_{\check{\Gamma}_w,G}$. Now at the generic point of $\check{\Gamma}_w$ the cycle $G_{p,q}$ is cut out on G by $S_{p,q}$ and hence the local ring of $\check{\Gamma}_w$ on $G_{p,q}$ is of the form $\mathcal{O}_{\check{\Gamma}_w,G}/(\lambda x^p - y^q)$. Henceforth the proof can be completed as in the proof of 19.4.(b). □

19.11. Lemma

The cycles Γ_P and $\Gamma_{w,P}$ are not contained in $G_{p,q}$.

Proof

Since $\check{\Gamma}_w$ is a component of $G_{p,q}.L_{Q,Q'}$, we may write

$$G_{p,q}.L_{Q.Q'} = \rho\check{\Gamma}_w + \Gamma ,$$

where Γ is an effective cycle not containing $\check{\Gamma}_w$ as a component. We are going to see that $\rho = p$ and $\Gamma \sim (2p+q,\ 2p+2q)^T$. In order to do that we compute $G.S_{p,q}.L_{Q,Q'}$ in two ways: on the one hand

$$\begin{aligned} G.S_{p,q}.L_{Q,Q'} &= (T_{P,Q,Q'}+\check{D}_w).S_{p,q} && \text{(by 19.8(a))} \\ &= T_{P,Q,Q'}.S_{p,q} + p\check{\Gamma}_w && \text{(by 15.3(b))} \end{aligned}$$

and on the other

$$\begin{aligned} G.S_{p,q}.L_{Q,Q'} &= G_{p,q}.L_{Q,Q'} + (2p+q)D_P.L_{Q,Q'} + q\check{A}_{w,P}.L_{Q,Q'} \\ &= \rho\check{\Gamma}_w + \Gamma + (2p+q)\Gamma_{P;v,v'} + q\check{\Gamma}_{v,P,Q} + q\check{\Gamma}_{v',P,Q'} . \end{aligned}$$

Comparing both expressions we get $\rho = p$ and that

$$G_{p,q}.L_{Q,Q'} = p\check{\Gamma}_w + \Gamma$$

$$(*) \qquad T_{P,Q,Q'}.S_{p,q} = \Gamma + (2p+q)\Gamma_{P,v,v'} + q\check{\Gamma}_{v,P,Q} + q\,\check{\Gamma}_{v',P,Q'} .$$

Since $G_{p,q}.L_{Q,Q'} \sim \begin{pmatrix} 2p+q \\ 4p+2q \end{pmatrix}$, one computes that $\Gamma \sim \begin{pmatrix} 2p+q \\ 2p+2q \end{pmatrix}$. So $\Gamma.\check{A} = 2p$. Moreover, the relation $S_{p,q}.\check{\Gamma}_{w,Q,Q'} = (2q,2p)\cdot\begin{pmatrix} 0 \\ 1 \end{pmatrix} = 2p$, together with (*), imply that the 2p intersections of Γ with $\check{A}$ actually lie on $\check{\Gamma}_{w,Q,Q'}$. In particular

$$(**) \qquad \Gamma \cap \check{\Gamma}_{v,P,Q} = \emptyset \quad \text{and} \quad \Gamma \cap \check{\Gamma}_{v',P,Q'} = \emptyset .$$

Assume now that $\Gamma_P \subseteq G_{p,q}$. Then $\Gamma_P \cap L_{Q,Q'} = \{(v^2,P^2),(v'^2,P^2)\}$ would be contained in $G_{p,q} \cap L_{Q,Q'} = \Gamma \cup \check{\Gamma}_w$. But this is absurd, because by (**) it cannot be contained in Γ, and clearly it is not contained in

$\check{\Gamma}_W$.

A similar argument shows that $\Gamma_{W,P}$ is not contained in $G_{p,q}$. In this case $\Gamma_{W,P} \cap L_{Q,Q'} = \{(v^2,Q^2),(v'^2,Q'^2)\}$ and this is not contained in $\check{\Gamma}_W$, nor in Γ. □

Construction of the codimension four cycles $\Gamma_{p,q}$

Here we explain a procedure for constructing a 1-dimensional system of conics havint (p,q) as its only pair of local characteristic numbers, p,q coprime positive integers.

Let P_o, P_1, P_2 be a triangle in $\mathbb{P}_2$ and let $T = T_{P_o,P_1,P_2}$. Then we know that the image of the restriction of $p\colon \mathbf{W} \longrightarrow \mathbb{P}_5$ to T is the plane T' spanned by u_o^2, u_1^2, u_2^2, where u_o, u_1, u_2 are the sides of the triangle P_o, P_1, P_2 and that $p\colon T \longrightarrow T'$ is the blowing up of T' at the three points u_o^2, u_1^2, u_2^2.

The plane T' can be easily described if we take P_o, P_1, P_2 as coordinate triangle: if this is the case the point conics in T' are those having a diagonal matrix $\mathrm{diag}(a_o,a_1,a_2)$. Moreover a_o,a_1,a_2 are projective coordinates on T' with respect to the triangle u_o^2, u_1^2, u_2^2.

Now given a pair (p,q) of coprime positive integers, let $\Gamma'_{p,q}$ be the curve in T' defined by the equation

$$\Gamma'_{p,q} : \quad a_o^p(a_1+a_2)^q = a_1^{p+q} \quad .$$

This curve has a single branch at u_o^2 of multiplicity q whose tangent is distinct from the sides of the triangle; it does not go through u_1^2; and it has a single branch at u_2^2 of multiplicity p and class q whose tangent coincides with the side $a_o=0$. On the side $a_2=0$ has p additional points, other that u_o^2.

Define $\Gamma_{p,q}$ as the strict transform of $\Gamma'_{p,q}$ under $p: T \longrightarrow T'$. Thus $\Gamma_{p,q}$ is the 1-dimensional system of conics which is the clousure of the non-degenerate conics that have P_o, P_1, P_2 as a self-polar triangle and satisfy the equation above.

From the standard properties of the blowing up one sees (cf. also §10) that $\Gamma_{p,q}$ has just one point on B, namely the point on the exceptional line over u_2^2 corresponding to the tangent $a_o = 0$, (u_2^2, P_o^2). Since the intersection numbers of $\Gamma_{p,q}$ with $\check{A}$ and A at this point are p and q, respectively, we see that the multiplicity cycle of $\Gamma_{p,q}$ at B is (u_2^2, P_o^2), and that it has a single characteristic pair, (p,q).

Finally $\Gamma_{p,q} \sim \binom{p+q}{p+q}$, as $\Gamma_{p,q} \cdot A = \Gamma_{p,q} \cdot \check{A} = p+q$, as one sees immediately from the description of $\Gamma'_{p,q}$ given before.

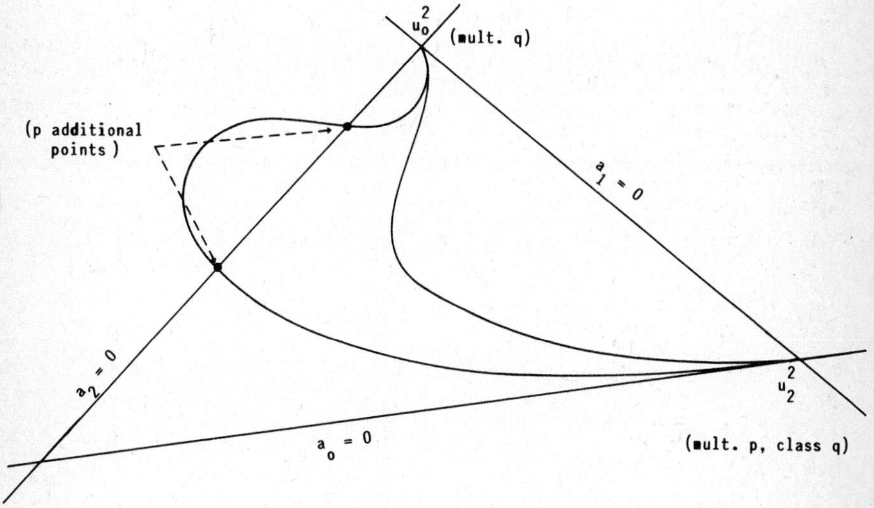

§ 20. Strict equivalence of conditions

20.1. Definition

Given degeneration free conditions K_1, K_2 of the same dimension d we shall say that K_1 is *strictly equivalent* to K_2 iff for any condition K of dimension 5-d the relation $p(K_1 \cdot K) = p(K_2 \cdot K)$ holds. □

In this section we will prove that two conditions are strictly equivalent iff they have the same global and local characters and then we will use this fact to construct free **Z**-basis of $Hal^i(W)$, for $i=2,3,4$. Given a condition K, its class $[K]$ in $A^{\cdot}(W)$ will be called *global character* of K. Identifying $A^i(W)$ with $\mathbf{Z}^{r(i)}$, where $r(0) = r(5) = 1$, $r(1) = r(4) = 2$, and $r(2) = r(3) = 3$, by means of the bases described in § 15, the global character of a codimension i condition is a vector with $r(i)$ integer coordinates. On the other hand, if K is a condition of codimension i we shall say that the pair (p,q) of coprime positive integers appears with multiplicity $\beta \in A^{i-1}(B)$ if β is the class of $K_{p,q}$ in $A^{i-1}(B)$. If we identify $A^j(B)$ with $\mathbf{Z}^{r'(j)}$, where $r'(0) = r'(3) = 1$, and $r'(1) = r'(2) = 2$, using the bases described in § 15, then the *multiplicity* of (p,q) in K is a vector with $r'(i-1)$ coordinates, $1 \leqslant i \leqslant 4$. With these conventions, the symbol $\beta[p,q]$ will be called the *local character* of K corresponding to the pair (p,q). The strict class of K will be denoted by $\langle K \rangle$.

Halphen's formula for higher codimensions (theorem 18.5) can now be rephrased, for the case of conics, as follows:

20.2. Let K, K' be degeneration free conditions of dimensions i and 5-i, respectively. Let α, α' be their global characters and $\beta_i[p_i,q_i]$, $1 \leqslant i \leqslant s$, $\beta'_j[p'_j,q'_j]$, $1 \leqslant j \leqslant s'$, their local characters. Then

$$p(K.K') = \alpha . \alpha' - \sum_{i,j} (\beta_i \cdot \beta'_j) \min(p_i q'_j, q_i p'_j) \quad \square$$

20.3. In view of the additive behaviour of this formula one can extend to non-effective cycles that do not have degenerate components the notions of proper intersection, strict equivalence, and of global and local characters. With such an extension the formula is still correct. □

20.4. Theorem

Let K and K' be conditions of the same codimension, say i. Then K and K' are strictly equivalent iff they have the same global and local characters.

Before proving this result we will use it to construct bases for $\mathrm{Hal}^i(\mathbf{W})$, i=2,3,4.

Since the strict equivalence is clearly compatible with addition, the strict equivalence classes of degeneration free cycles of codimension i form an abelian group under addition. This group will be denoted by $\mathrm{Hal}^i(\mathbf{W})$, so that $\mathrm{Hal}^1(\mathbf{W})$ is the group $\mathrm{Hal}(\mathbf{W})$ studied in §12. The strict class of a condition K will be denoted by $\langle K \rangle$. Notice that $\mathrm{Hal}^o(\mathbf{W})$ is infinite cyclic generated by $\langle \mathbf{W} \rangle$ and that $\mathrm{Hal}^5(\mathbf{W})$ is infinite cyclic generated by $\langle C \rangle$, where C is any point of **W**.

With the notations of § 12, and fixing a point P, a line u, and all other points and lines needed to define the cycles $H_{p,q}$, $G_{p,q}$, and $\Gamma_{p,q}$ we have the following:

20.5. The estrict equivalence classes of the cycles

$$L_P^2 , \quad \frac{L_P \cdot \check{L}_u}{2} , \quad \check{L}_u^2 , \quad H_{p,q} , \quad \check{H}_{p,q} ,$$

where (p,q) runs through all pairs of coprime positive integers, form a free basis of $Hal^2(W)$. Moreover, given a condition K if we let $(a,b,\check{a}) \in \mathbf{Z}^3$ be its global character and $\beta_{p,q}$ $[p,q]$ its local characters, $\beta_{p,q} = (\beta^1_{p,q}, \beta^2_{p,q}) \in \mathbf{Z}^2$, then

$$<K> = (a - \Sigma\, q\beta^1_{p,q})<L^2> + (b - \Sigma(4\beta^1_{p,q}+\beta^2_{p,q})p - \Sigma(\beta^1_{p,q}+4\beta^2_{p,q})q) < \frac{L\check{L}}{2} > +$$

$$+ (\check{a} - \Sigma\, p\beta^2_{p,q})<\check{L}^2> + \Sigma\beta^1_{p,q}<H_{p,q}> + \Sigma\,\beta^2_{p,q} <\check{H}_{p,q}>.$$

20.6. Let S, T and $\check{S}$ be the basis of $A^3(W)$ that is dual of the basis L^2, $\frac{L\check{L}}{2}$, $\check{L}^2$ of $A^2(\mathbf{W})$. Then the strict classes of

$$S,\ T,\ \check{S},\ G_{p,q},\ \check{G}_{p,q}\ ,$$

where (p,q) runs through all pairs of coprime positive integers, form a free basis of $Hal^3(\mathbf{W})$. Furthermore, given a codimension 3 condition K if we let $(a,b,\check{a}) \in \mathbf{Z}^3$ denote its global character and $\beta_{p,q}[p,q]$ its local characters, $\beta_{p,q} = (\beta^1_{p,q}, \beta^2_{p,q}) \in \mathbf{Z}^2$, then

$$\begin{aligned} < K > = {} & (a - \Sigma\,(2p+q)\beta^1_{p,q} - \Sigma\,(2p+2q)\beta^2_{p,q}) < S > \\ & + (b - \Sigma(2p+q)\beta^1_{p,q} - \Sigma\,(p+2q)\beta^2_{p,q}) < T > \\ & + (\check{a} - \Sigma(2p+2q)\beta^1_{p,q} - \Sigma(p+2q)\beta^2_{p,q}) < \check{S} > \\ & + \Sigma\,\beta^1_{p,q} < G_{p,q} > + \Sigma\,\beta^2_{p,q} < \check{G}_{p,q} >. \end{aligned}$$

20.7. Let Γ, $\check{\Gamma}$ be the basis of $A^4(\mathbf{W})$ that is dual of the basis L, $\check{L}$ of $A^1(\mathbf{W})$. Then the strict classes of

$$\Gamma\ ,\ \check{\Gamma}\ ,\ \Gamma_{p,q}\ ,$$

where (p,q) runs over all pairs of coprime positive integers, form a free basis of $Hal^4(\mathbf{W})$. Moreover, if K is a codimension 4 condition, and we let $(a,\check{a})$ be its global character and $\beta_{p,q}[p,q]$, $\beta_{p,q} \in \mathbf{Z}$, its local characters, then

$$<K> = (a - \Sigma(p+q)\beta_{p,q})<\Gamma> + (\check{a} - \Sigma(p+q)\beta_{p,q})<\check{\Gamma}> + \Sigma\ \beta_{p,q} <\Gamma_{p,q}>.$$

Proofs

In the three cases the expression on the right hand side has the same characters as K, as a straightforward computation shows. Therefore the claimed equalities are valid by 20.4. So the system of conditions in each of the three cases is a system of generators of the corresponding $Hal^i(\mathbf{W})$. Now, using 20.4, it is easy to check that these generators are linearly independent over $\mathbb{Z}$. □

Proof of 20.4.

That the conditions are sufficient is an inmediate consequence of Halphen's generalized formula 20.2.

To see that the conditions are necessary, assume K and K' are strictly equivalent. Intersect first K and K' with the cycles L^{5-i}, $L^{4-i}\check{L}$, ..., $\check{L}^{5-i}$. Since these cycles do not have local characteristics, the intersection numbers are the same (they coincide with the proper intersection numbers) and we deduce easily that K and K' have the same global characters. We want to see that they also have the same local characters.

Let $(p_1,q_1),\ldots,(p_\ell,q_\ell)$ be any set of pairs of coprime positive integers that contains all pairs of local characteristic numbers of K and K'. For $1 \leqslant j \leqslant \ell$, let β_j denote the multiplicity of (p_j,q_j) in K if i=1 or 4, and either the first or the second component of the multiplicity of (p_j,q_j) in K if i=2 or 3. Define β_j' similarly for K'. We want to see that $\beta_j = \beta_j'$. Without loss of generality we may assume that p_j/q_j increases with j.

In order to do that, set, for $1 \leqslant j \leqslant \ell$, $K_j = S_{p_j,q_j}$ if i=1; $K_j =$

$= H_{p_j,q_j}$ or $\check{H}_{p_j,q_j}$ if $i=2$ and according to whether we are looking at the first or second component of the multiplicity; $K_j = G_{p_j,q_j}$ or $\check{G}_{p_j,q_j}$ if $i=3$ and according to whether we are looking at the first or second component of the multiplicity; and $K_j = \Gamma_{p_j,q_j}$ if $i=4$. In all cases, if we set $\pi_j := p(K,K_j)$, then by 20.2 we get that

$$\pi_j = (\sum_{h=1}^{j} \beta_h \, p_h) q_j + (\sum_{h=j+1}^{\ell} \beta_h \, q_h) p_j \ .$$

From this expression it is immediate to deduce that

$$q_j \, \pi_{j+1} - q_{j+1} \pi_j = (q_j p_{j+1} - p_j q_{j+1})(\sum_{h=j+1}^{\ell} \beta_h \, q_h)$$

and

$$p_{j+1} \pi_j - p_j \, \pi_{j+1} = (q_j p_{j+1} - p_j q_{j+1})(\sum_{h=1}^{j} \beta_h \, p_h) \ .$$

So if we set, for $1 \leq j < \ell$,

$$\Delta_j = q_j p_{j+1} - p_j q_{j+1}, \quad \Delta_j' = q_j \, \pi_{j+1} - q_{j+1} \pi_j, \quad \Delta_j'' = p_{j+1} \pi_j - p_j \pi_{j+1}$$

and

$$\rho_j' = \Delta_j'/\Delta_j \quad , \quad \rho_j'' = \Delta_j''/\Delta_j$$

then

$$\rho_j' = \beta_{j+1} q_{j+1} + \cdots + \beta_\ell \, q_\ell$$

and

$$\rho_j'' = \beta_1 p_1 + \cdots + \beta_j p_j \ .$$

It turns out that for $1 < j < \ell-1$

(*) $$\rho_{j-1}' - \rho_j' = \beta_j q_j \quad \text{and} \quad \rho_j'' - \rho_{j-1}'' = \beta_j p_j$$

whereas

(**) $$\rho_{\ell-1}' = \beta_\ell \, q_\ell \quad \text{and} \quad \rho_1'' = \beta_1 p_1 \ .$$

Since the numbers ρ_j' and ρ_j'' only depend on the sequence $\pi_1,\dots,\pi_\ell$, these expressions show that the multiplicites β_j only depend on the sequence $\pi_1,\dots,\pi_\ell$. But since K and K' are strictly equivalent, $p(K,K_j) = p(K',K_j)$ and so the above expressions (*) and (**) show that $\beta_j = \beta_j'$ for all j. This completes the proof of 20.4. □

In next section we need the fact that the K_j's used in the proof above have some sort of universality that we explain presently.

20.8. Proposition

Given a class $\xi \in A^i(\mathbf{W})$, there exist effective cycles $K_1,\dots,K_\ell$ of codimension 5-i, depending only on ξ, such that the local characters of any effective cycle K representing ξ can be computed from the set of integers $\pi_j = p(K,K_j)$, $1 \leqslant j \leqslant \ell$. (The actual expressions are as (*) and (**) in the preceding proof.)

Proof

By the proposition 20.10 below, the local characters of an effective K representing ξ have an upper bound that only depends on ξ. Then if $(p_1,q_1),\dots,(p_\ell,q_\ell)$ is the set of all pairs of coprime positive integers satisfying this bound, any effective cycle K representing ξ only will have characteristic pairs belonging to this set and so the cycles K_j constructed as in the proof of 20.4 only depend on ξ. From this the proof of the proposition follows readily. □

The rest of the section is devoted to prove proposition 20.10. We will first prove a lemma.

20.9. Lemma

Let K be an effective degeneration free cycle on $\mathbf{W}$, Z and excedentary component of $K \cap \mathbf{B}$, and (p,q) a local characteristic pair of K at

Z. Let $M \subseteq \mathbf{A}$ be a cycle such that $\dim(M)+\dim(K) = 5$ and the intersection $K \cap M$ is proper (i.e., 0-dimensional). Then for any $C \in M \cap Z$ we have that

$$i_C(\alpha_*(M).K) \geq q ,$$

where $\alpha : \mathbf{A} \longrightarrow \mathbf{W}$ is the inclusion map.

Similarly, if $M' \subseteq \check{\mathbf{A}}$ is such that the intersection $K \cap M'$ is proper and $\dim(M') = 5-\dim(M)$, then for any $C' \in M' \cap Z$ we have that

$$i_{C'}(\check{\alpha}_*(M').K) \geq p ,$$

where $\check{\alpha} : \check{\mathbf{A}} \longrightarrow \mathbf{W}$ is the inclusion map.

Proof

Since the intersections $M \cap \alpha^*K$ and $\alpha_*(M) \cap K$ are proper on $\mathbf{A}$ and $\mathbf{W}$, respectively, by the projection formula we have that

$$\alpha_*(M.\alpha^*K) = \alpha_*M.K$$

Now Z is a component of α^*K whose multiplicity in α^*K is $\mu = i_Z(\mathbf{A}.K)$, and so from the equality above we see that

$$\begin{aligned} i_C(\alpha_*M.K) &= i_C(M.\alpha^*K) \\ &\geq i_C(M.\mu K) \\ &\geq \mu , \end{aligned}$$

and so it is enough to see that $\mu \geq q$. But by the definition of local characteristic numbers there exists a discrete valuation v of $\mathcal{O}_{K,Z}$ such that $v(x) = rq$, $r \geq 1$, where x is a generator of the ideal of $\mathbf{A}$ in $\mathcal{O}_{K,Z}$. On the other hand μ is the order of x in $\mathcal{O}_{K,Z}$, because $\mathbf{A}$ is a smooth hypersurface, and so $\mu \geq rq$, for instance by [F], A.3.1. So $\mu \geq q$.

The last statement is proved similarly. □

20.10. Proposition.

Given an effective codimension i degeneration free cycle K on $\mathbf{W}$ there exists an upper bound of the local characteristic numbers of K that depends only on the class $[K] \in A^i(\mathbf{W})$.

Proof

For codimension 1 cycles it is a consequence of corollary 10.4, while for codimension 4 cycles it is a consequence of remark 3.5. So we only need consider the cases i=2 and i=3. In both cases there exists an effective cycle M on $\mathbf{A}$, of codimension 5-i, whose restriction to $\mathbf{B}$ has positive intersection number with any effective cycle on $\mathbf{B}$. In fact it clearly suffices to take for M an effective cycle on $\mathbf{A}$ whose restriction to $\mathbf{B}$ is $\ell + \check{\ell}$ for i=3 and $\ell^2 + \check{\ell}^2$ for i=2. If this is the case, if Z is an excedentary component of $K \cap \mathbf{B}$, and if (p,q) is a pair of local characteristic numbers of K at Z, then for $\sigma \in G$ generic $\sigma(M)$ intersects Z properly and by construction of M, $\sigma(M) \cap Z$ is non-empty. Therefore, by lemma 20.9,

$$q \leqslant (\alpha_*(\sigma(M)).K) ,$$

and $\alpha_*(\sigma(M)).K$ only depends on the global character $[K]$ of K. The proof that p is bounded likewise is done similarly, using $\check{\mathbf{A}}$ instead of $\mathbf{A}$. □

§ 21. Enumerative ring of W

Let $\mathrm{Hal}^{\bullet}(\mathbf{W})$ be the direct sum of the groups $\mathrm{Hal}^i(\mathbf{W})$ defined in last section. Then we have a non-degenerate pairing

$$\mathrm{Hal}^i(\mathbf{W}) \times \mathrm{Hal}^{5-i}(\mathbf{W}) \longrightarrow \mathbf{Z}$$

induced by the proper intersection pairing. Now $\mathrm{Hal}^5(\mathbf{W}) \xrightarrow{\sim} \mathbf{Z}$ and so we can regard this pairing as a pairing

$$\mathrm{Hal}^{i}(\mathbf{W}) \times \mathrm{Hal}^{j}(\mathbf{W}) \longrightarrow \mathrm{Hal}^{5}(\mathbf{W})$$

defined for $i+j = 5$.

The goal of this section is to define pairings

$$\mathrm{Hal}^{i}(\mathbf{W}) \times \mathrm{Hal}^{j}(\mathbf{W}) \longrightarrow \mathrm{Hal}^{i+j}(\mathbf{W})$$

for all i,j that extend the pairing above. It turns out that $\mathrm{Hal}^{\bullet}(\mathbf{W})$, with the product that these pairings define, is a commutative graded ring with unit (theorem 21.6). The enumerative significance of this product is the contents of theorem 21.7. This ring will be called *Halphen's ring* (or *strict intersection ring* , or *enumerative ring*) of **W**. This ring coincides with that introduced by De Concini and Procesi in [DC-P] for the case of non-degenerate conics acted upon by PGL(3).

To start with, suppose that K and K' are irreducible cycles of codimensions i and j, respectively, such that their traces K_o and K'_o on $\mathbf{W}_o := \mathbf{W} - \mathbf{A} - \check{\mathbf{A}}$ intersect properly on $\mathbf{W}_o$. Then we will set $K \cap K'$ to denote the clousure in **W** of the intersection cycle $K_o.K'_o$. We need a few properties of this cycle that we explain in two lemmas.

21.1. Lemma

Let K_1, K_2, K_3 be irreducible cycles of codimensions i_1, i_2, i_3 and assume $i_1+i_2+i_3 = 5$. Then there exists non-empty open sets U and U' of G such that

$$p(K_1 \cap K_2^{\sigma}, K_3) = p(K_1, K_2^{\sigma'} \cap K_3)$$

for all $\sigma \in U$, $\sigma' \in U'$. Therefore the common value of these expressions is independent of $\sigma \in U$ and $\sigma' \in U'$.

Proof

Let U_1 be a non-empty open set of G^2 such that for $(\sigma,\sigma_1) \in U_1$ the intersection $K_1 \cap K_2^{\sigma} \cap K_3^{\sigma_1} \cap W_o$ has finitely many points, each counted with multiplicity one, and so that $\# K_1 \cap K_2^{\sigma} \cap K_3^{\sigma_1} \cap W_o$ is independent of $(\sigma,\sigma_1) \in U_1$ (use an argument similar to the argument explained in the proof of 14.1). Let $u: G^2 \longrightarrow G$ be the projection onto the first factor. Let $V \subset G$ be a non-empty open set such that $K_1 \cap K_2^{\sigma} \cap W_o$ is reduced and of codimension i_1+i_2, and set $U_2 := U_1 \cap u^{-1}(V)$. Then U_2 is non-empty and we claim that

$$(*) \qquad p(K_1 \cap K_2^{\sigma}, K_3) = \#(K_1 \cap K_2^{\sigma} \cap K_3^{\sigma_1} \cap W_o)$$

for all $(\sigma,\sigma_1) \in U_2$.

In order to prove the claim, take any $\sigma \in u(U_2)$, so that $K_1 \cap K_2^{\sigma}$ is defined and reduced. Then by 17.1 there exists a non-empty open set $V_{\sigma} \subset G$ (depending on σ) such that for any $\sigma_1 \in V_{\sigma}$

$$\begin{aligned} p(K_1 \cap K_2^{\sigma}, K_3) &= \#(K_1 \cap K_2^{\sigma} \cap K_3^{\sigma_1} \cap W_o) \\ &= \#(K_1 \cap K_2^{\sigma} \cap K_3^{\sigma_1} \cap W_o) \quad , \end{aligned}$$

the last equality by definition of $K_1 \cap K_2^{\sigma}$. Now $U_2 \cap u^{-1}(\sigma)$ and $\{\sigma\} \times V_{\sigma}$ are non-empty open sets in $u^{-1}(\sigma)$ and so their intersection is non-empty. Moreover, the claim (*) holds for any $(\sigma,\sigma_1) \in U_2 \cap \{\sigma\} \times V_{\sigma}$. But since in (*) the left hand side is independent of σ_1, and so is the second for $(\sigma,\sigma_1) \in U_2$, the claim follows.

Likewise there exists a non-empty open set $U_2' \subset G^2$ such that for any $(\sigma',\sigma_1') \in U_2$

$$(**) \qquad p(K_1^{\sigma_1'}, K_2^{\sigma'} \cap K_3) = \#(K_1^{\sigma_1'} \cap K_2^{\sigma'} \cap K_3 \cap W_o) \quad ,$$

and so that the common value is independent of $(\sigma',\sigma_1') \in U_2'$. Now we observe that

$$\#(K_1 \cap K_2^{\sigma} \cap K_3^{\sigma_1} \cap W_o) = \#(K_1^{\sigma_1'} \cap K_2^{\sigma'} \cap K_3 \cap W_o)$$

for any $(\sigma,\sigma_1) \in U_2$, $(\sigma',\sigma_1') \in U_2'$. To see this it is enough to show there exists $(\sigma,\sigma_1) \in U_2$ and $(\sigma',\sigma_1') \in U_2'$ for which it is true. But this is accomplished taking any $(\sigma,\sigma_1) \in U_2$ such that $(\sigma_1^{-1}\sigma, \sigma_1^{-1}) \in U_2'$, which can clearly be done.

To end the proof of the lemma, it is enough to take non-empty open sets U, U' of G that are contained in $u(U_2)$ and in $u(U_2')$, respectively. □

21.2. Lemma

Let K_1, K_2 be irreducible cycles. Then there exists a non-empty open set U of G such that the cycle $K_1 \pitchfork K_2^{\sigma}$ is defined for any $\sigma \in U$ and so that its strict equivalence class does not depend on $\sigma \in U$. Moreover, if V is the open set of G^2 of those (σ',σ'') such that $(\sigma')^{-1}\sigma'' \in U$, then $K_1^{\sigma'} \pitchfork K_2^{\sigma''}$ is defined for any $(\sigma',\sigma'') \in V$ and

$$< K_1^{\sigma'} \pitchfork K_2^{\sigma''} > = < K_1 \pitchfork K_2^{\sigma} >,$$

where $\sigma = (\sigma')^{-1}\sigma''$, so that $< K_1^{\sigma'} \pitchfork K_2^{\sigma''} >$ is independent of $(\sigma',\sigma'') \in V$.

Proof

Given a cycle K_3 with $cod(K_1)+cod(K_2)+cod(K_3) = 5$ there exists, by 21.1, a non-empty open set U_{K_3} of G such that $p(K_1 \pitchfork K_2^{\sigma}, K_3)$ is independent of $\sigma \in U_{K_3}$. Taking for K_3 conditions of the form $L^i\check{L}^j$ we see that there exists a non-empty open set U of G such that the global characters of $K_1 \pitchfork K_2^{\sigma}$ are independent of σ. Now we use 20.4, 20.8 and 21.1 again, to conclude that there exists a non-empty open set U' of G such that the global and local characters of $K_1 \pitchfork K_2^{\sigma}$ are independent of $\sigma \in U'$. This proves the first assertion.

For the second assertion notice that

$$K_1^{\sigma'} \cap K_2^{\sigma''} = (K_1 \cap K_2^{\sigma})^{\sigma'} ,$$

with $\sigma = (\sigma')^{-1} \circ \sigma''$, so that indeed $K_1^{\sigma'} \cap K_2^{\sigma''}$ is defined and its strict class is $< K_1 \cap K_2^{\sigma} >$, which by the first part does not depend on $\sigma \in U$. □

21.3. Remark

For convenience of the proofs, lemmas 21.1 and 21.2 are stated for irreducible cycles, by they can be easily extended to arbitrary cycles. □

In order to define a product in $Hal^{\bullet}(\mathbf{W})$, we first define biadditive maps

$$Z^i(\mathbf{W}) \times Z^j(\mathbf{W}) \xrightarrow{\quad p \quad} Hal^{i+j}(\mathbf{W})$$

by the formula

$$p(K_1,K_2) = < K_1 \cap K_2^{\sigma} > ,$$

where $\sigma \in G$ is chosen generically. That this is well defined is the contents of the first assertion of lemma 21.2 (plus remark 21.3).

Notice that if $i+j = 5$ then $p(K_1,K_2)$ coincides with the proper intersection number.

21.4. The map p is symmetrical, i.e., $p(K_2,K_1) = p(K_1,K_2)$.

Proof

It is an immediate consequence of lemma 21.2. □

Next step is to see that p factors through strict equivalence.

21.5. Lemma

Assume K_1 and K_1' are strictly equivalent cycles. Then

$$p(K_1,K_2) = p(K_1',K_2)$$

for any cycle K_2.

Proof

Let $U \subseteq G$ be a non-empty open set such that, for $\sigma \in U$,

$$p(K_1,K_2) = \langle K_1 \pitchfork K_2^\sigma \rangle, \quad p(K_1',K_2) = \langle K_1' \pitchfork K_2^\sigma \rangle .$$

Then we want to see that $K_1 \pitchfork K_2^\sigma$ and $K_1' \pitchfork K_2^\sigma$ are strictly equivalent for $\sigma \in U$. In order to see this pick any cycle K_3 of codimension $i_3 := 5-i_1-i_2$, where i_1, i_2 are the codimensions of K_1, K_2, respectively. By 21.1 (and remark 21.3) there exists a non-empty open set U' of G such that $K_1 \pitchfork K_2^\sigma$, $K_1' \pitchfork K_2^\sigma$, $K_2^\sigma \pitchfork K_3$ are defined and

$$\begin{aligned} p(K_1 \pitchfork K_2^\sigma, K_3) &= p(K_1, K_2^\sigma \pitchfork K_3), \\ p(K_1' \pitchfork K_2^\sigma, K_3) &= p(K_1', K_2^\sigma \pitchfork K_3), \end{aligned}$$

for any $\sigma \in U'$. Now by definition of the strict equivalence we have that

$$p(K_1, K_2^\sigma \pitchfork K_3) = p(K_1', K_2^\sigma \pitchfork K_3),$$

so that, for any $\sigma \in U'$,

$$p(K_1 \pitchfork K_2^\sigma, K_3) = p(K_1' \pitchfork K_2^\sigma, K_3) .$$

But since $\langle K_1 \pitchfork K_2^\sigma \rangle$ and $\langle K_1' \pitchfork K_2^\sigma \rangle$ are constant for $\sigma \in U$, this equality holds for $\sigma \in U$ and therefore $K_1 \pitchfork K_2^\sigma$ and $K_1' \pitchfork K_2^\sigma$ are strictly equivalent. □

From 21.6 and 21.5 it follows that the value $P(K_1,K_2)$ only depends on the strict equivalence class of K_1 and K_2, so that there exists a unique biadditive map

$$\mathrm{Hal}^i(\mathbf{W}) \times \mathrm{Hal}^j(\mathbf{W}) \longrightarrow \mathrm{Hal}^{i+j}(\mathbf{W})$$

such that

$$(<K_1>, <K_2>) \longrightarrow p(K_1,K_2) .$$

The value of this map on the pair of classes (α,β) will be denoted by $\alpha \cdot \beta$ and will say that it is the Halphen product of α and β. When $i+j = 5$ this product is the proper intersection number of corresponding representatives of α and β.

21.6. Theorem

$\mathrm{Hal}^{\cdot}(\mathbf{W})$, equipped with the Halphen product, is a commutative graded ring with unit.

Proof

The class of $\mathbf{W}$ is clearly a unit element for the Halphen product. The commutativity follows from 21.4. The only property that now needs to be checked is the associativity.

The associativity will be proved by means of the following lemma:

> Given irreducible cycles K_1,K_2,K_3 there exists a non-empty open set $U \subseteq G^2$ such that the intersection $K_1 \cap K_2^{\sigma'} \cap K_3^{\sigma''} \cap \mathbf{W}_o$ is proper on $\mathbf{W}_o$ and the strict class
>
> $$\alpha = \overline{<K_1 \cap K_2^{\sigma'} \cap K_3^{\sigma''} \cap \mathbf{W}_o>}$$
>
> is independent of (σ',σ'') provided $(\sigma',\sigma'') \in U$.

First let us show that this statement implies associativity. In order to prove this property it is enough to see that if K_1,K_2,K_3 are irreducible cycles then

$$(<K_1>.<K_2>).<K_3> = <K_1>.(<K_2>.<K_3>) .$$

To see this, let U be an open set of G^2 as in the lemma and let $\pi: G^2 \longrightarrow G$ be the projection onto the first factor. Take $V \subseteq \pi(U)$, V open and non-empty, and such that $K_1 \pitchfork K_2^{\sigma}$ is defined for $\sigma \in V$ and its strict class is independent of $\sigma \in V$. For each $\sigma \in V$ let V_σ be a non-empty open set of G such that $\sigma \times V_\sigma \subseteq U$ and in such a way that $(K_1 \pitchfork K_2^{\sigma}) \pitchfork K_3^{\sigma'}$ is defined for any $\sigma' \in V_\sigma$ and its strict class independent of $\sigma' \in V_\sigma$.

By definition

$$(<K_1>.<K_2>).<K_3> \quad = \quad <(K_1 \pitchfork K_2^{\sigma}) \pitchfork K_3^{\sigma'}>$$

provided $\sigma \in V$ and $\sigma' \in V_\sigma$. Since $(\sigma, \sigma') \in U$, $<(K_1 \pitchfork K_2^{\sigma}) \pitchfork K_3^{\sigma'}> = \alpha$, as it is easy to see by the associativity of the intersection product on $\mathbf{W}_o$.

On the other hand there exists a non-empty open set $\tilde{V}$ of G, and for each $\sigma \in \tilde{V}$ a non-empty open set $\tilde{V}_\sigma$ of G such that for all $\sigma \in \tilde{V}$, $\sigma' \in \tilde{V}_\sigma$

$$<K_1>.(<K_2>.<K_3>) = <K_1 \pitchfork (K_2 \pitchfork K_3^{\sigma})^{\sigma'}>$$

and with $(\sigma', \sigma'\sigma) \in U$. Then since

$$<K_1 \pitchfork (K_2 \pitchfork K_3^{\sigma})^{\sigma'}> \; = \; <\overline{K_1 \cap (K_2^{\sigma'} \cap K_3^{\sigma'\sigma}) \cap \mathbf{W}_o}>$$

we see that

$$<K_1>.(<K_2>.<K_3>) \; = \; \alpha ,$$

and this proves the associativity.

It remains to prove the lemma. In order to do that, let K_4 be any cycle such that $\sum_1^4 \operatorname{cod}(K_i) = 5$. Let $V \subseteq G^3$ be a non-empty open set such that $\#(K_1 \cap K_2^{\sigma_1} \cap K_3^{\sigma_2} \cap K_4^{\sigma_3} \cap \mathbf{W}_o) = \ell$ (say) is finite and independent of $\sigma = (\sigma_1, \sigma_2, \sigma_3) \in V$. Now we choose a non-empty open set U' in G^2 such that

(a) $K_1 \cap K_2^{\sigma'} \cap K_3^{\sigma''} \cap W_o$ is proper and reduced for $(\sigma',\sigma'') \in U'$, and

(b) U' is contained in the image of V under the projection of G^3 onto the first two factors.

With this, and arguing as in the proof of 21.1, one can easily see that

$$\overline{p(K_1 \cap K_2^{\sigma'} \cap K_3^{\sigma''} \cap W_o,\ K_4)} = \ell$$

for $(\sigma',\sigma'') \in U'$. Finally the lemma follows from the fact that only finitely many cycles K_4 are needed to determine the strict class of $K_1 \cap K_2^{\sigma'} \cap K_3^{\sigma''} \cap W_o$. This ends the proof of the theorem.

21.7. Theorem

Given reduced degeneration free conditions K_i, $i=1,\ldots,r$, with $\sum_1^r \operatorname{cod}(K_i) = 5$, there exists a non-empty open set $U \subseteq G^{r-1}$ such that for $\sigma = (\sigma_2,\ldots,\sigma_r) \in U$ the conics properly satisfying K_1, $\sigma_2(K_2), \ldots, \sigma_r(K_r)$ are non-degenerate, finite in number, and such that this number is constant and given by $\langle K_1\rangle \ldots \langle K_r\rangle$.

Proof

That there exists a non-empty open set U' of G^{r-1} such that for $\sigma = (\sigma_2,\ldots,\sigma_r) \in U'$ the conics properly satisfying K_1, $\sigma_2(K_2), \ldots, \sigma_r(K_r)$ are non-degenerate, finite in number, and counting with multiplicity one in the intersection of K_1, $\sigma_2(K_2), \ldots, \sigma_r(K_r)$ is an easy consequence of Kleiman's theorem and the definition of conic properly satisfying a condition. Therefore for $\sigma \in U'$ the set of conics properly satisfying K_1, $\sigma_2(K_2), \ldots, \sigma_r(K_r)$ is equal to

$$K_1 \cap \sigma_2(K_2) \cap \ldots \cap \sigma_r(K_r) \cap W_o \ .$$

Let U be a non-empty open subset of U' such that the number

n of elements of this set is constant for $\sigma \in U$. Now we will show that $n = \langle K_1\rangle \ldots \langle K_r\rangle$. Indeed, by definition of Halphen's product

$$\langle K_1\rangle \ldots \langle K_r\rangle = [(K_1 \Cap \sigma_2(K_2))\ldots] \Cap \sigma_r(K_r) \quad ,$$

where σ_2 may be chosen in a non-empty open set V of G and in general for each choice of $\sigma_2,\ldots,\sigma_{i-1}$ the σ_i may be chosen in a non-empty open set $V_{\sigma_2,\ldots,\sigma_{i-1}}$ of G depending on $\sigma_2,\ldots,\sigma_{i-1}$. Since we may chose the open sets V and $V_{\sigma_2,\ldots,\sigma_{i-1}}$ in such a way that the 0-cycle

$$[(K_1 \Cap \sigma_2(K_2))\ldots] \Cap \sigma_r(K_r)$$

is reduced, and that this cycle is clearly equal to

$$K_1 \cap \sigma_2(K_2) \cap \ldots \cap \sigma_r(K_r) \cap \mathbf{W}_o \quad ,$$

we only need to show that there exists $\sigma = (\sigma_2,\ldots,\sigma_r) \in G^{r-1}$ such that $\sigma \in U$ and $\sigma_i \in V_{\sigma_2,\ldots,\sigma_{i-1}}$. This can be done choosing σ_i in $V_{\sigma_2,\ldots,\sigma_{i-1}}$ and in the projection of $U \cap \{\sigma_2\} \times \ldots \times \{\sigma_{i-1}\} \times G^{r-i}$ in the i-th factor of $\{\sigma_2\} \times \ldots \times \{\sigma_{i-1}\} \times G^{r-i}$.

§ 22. Computation of products in Hal$^\bullet$(W)

Granted that we know how to express the strict class of any degeneration free condition as a linear combination of the free $\mathbb{Z}$-basis of Hal$^\bullet$($\mathbf{W}$) constructed in §§ 12 and 20 (which henceforth will be called the basis and its elements basis elements), the explicit solution to any enumerative problem of conics depends on the ability to compute the Halphen product of any two elements of that basis. Indeed, by theorem 21.7 the number of conics properly satisfying given conditions $K_1,\ldots,K_r$, with $\sum_1^r \mathrm{cod}(K_i) = 5$ (where the data of these conditions is in general position) is $\langle K_1\rangle\ldots\langle K_r\rangle$, and we will know how to evaluate this product once we know how to calculate the product of any two elements of the basis.

The computations will consist in determining the global and local characters of each product, inasmuch as this information is all we need to determine a strict class.

We begin with two lemmas. We will let u denote a line, P a point on u, and $S_{p,q}$ the condition defined using u and P as in (11.1).

22.1. Lemma (generalization of 13.1)

Let K be a degeneration free cycle on $\mathbf{W}$ such that $K \subseteq S_{p,q}$. Let Z be an excedentary component of $K \cap B$ and (p',q') a pair of local characteristic numbers of K at Z. Then

(a) If $p'/q' > p/q$, all conics in Z pass through P.

(b) If $p'/q' < p/q$, all conics in Z are tangent to u.

(c) Therefore $p'/q' = p/q$ if not all conics in Z go through P nor are all tangent to u.

Proof

Without loss of generality we may assume that K is irreducible. We know that $S_{p,q}$ lies in the pencil of divisors generated by $q\check{A} + (2p+q)\check{L}_u$ and $pA + (p+2q)L_P$ (see 11.1). Thus if $K \subseteq S_{p,q}$, then all divisors in the pencil other than $S_{p,q}$ cut out on K the same divisor. This implies that if $x, y, f, g \in \mathcal{O} := \mathcal{O}_{K,Z}$ are generators of the ideals of $A \cap K$, $\check{A} \cap K$, $\check{L}_u \cap K$, and $L_P \cap K$ in $\mathcal{O}$, respectively, then $x^p g^{p+2q}$ and $y^q f^{2p+q}$ generate the same ideal in $\mathcal{O}$, so that in particular

$$p\ v(x) + (p+2q)\ v(g) = q\ v(y) + (2p+q)\ v(f)$$

for any valuation of $\mathbb{C}(K)$ centered at $\mathcal{O}$. In particular this relation will hold true for the valuation v that defines the pair (p',q'). Dividing throughout by $q\,v(x)$ we get that

$$p/q + v(g)(p+2q)/qv(x) = p'/q' + v(f)(2p+q)/qv(x).$$

Hence, if $p'/q' < p/q$ then $v(f) > 0$, which means that $Z \subseteq L_u$, i.e., that all conics in Z are tangent to u. This proves (b). Similarly, dividing throughout by $pv(f)$ we deduce (a). □

22.2. Lemma

Let K be a degeneration free cycle on $\mathbf{W}$ without local characteristic numbers (i.e., $K \cap B$ is proper). Then there exists a non-empty open set $U \subseteq G$ such that for $\sigma \in U$

$$\sigma(K) \cap S_{p,q} = \sigma(K) \mathbin{\hat{\cap}} S_{p,q}$$

and $\bar{K}_\sigma = \sigma(K) \mathbin{\hat{\cap}} S_{p,q}$ has (p,q) as its only local characteristic pair. Moreover,

$$(\bar{K}_\sigma)_{p,q} = j^*(\sigma(K)) \quad ,$$

where $j\colon B \longrightarrow \mathbf{W}$ is the inclusion.

Proof

To prove the first assertion it is enough to see that there exists a non-empty open set $U \subseteq G$ such that $\sigma(K) \cap S_{p,q}$ does not have degenerate components. And this is clear, because of the hypothesis on K, by Kleiman's theorem.

Now we will see, possibly after shrinking U, that $\bar{K}_\sigma$ has, for $\sigma \in U$, only (p,q) as a local characteristic pair. In fact $\sigma(K) \cap S_{p,q} \cap B =$ $= \sigma(K) \cap B$ and if $\bar{K}_\sigma$ had a local characteristic pair $(p',q') \neq (p,q)$, then a component of $\sigma(K) \cap B$ would be contained either in $\check{L}_u$ or in L_P

(lemma 22.1). But since for σ generic this is impossible, because $\sigma(K)\cap B$ intersects properly $\check{L}_u \cap B$ and $L_p \cap B$ (for $\sigma \in G$ generic), the claimed shrinking of U is certainly possible.

It remains to compute $(\bar{K}_\sigma)_{p,q}$. From the definition of this cycle (18.1) it follows that $\bar{K}_\sigma . A$ is equal to $qj_*(\bar{K}_\sigma)_{p,q}$ plus components not contained in B. But

$$\begin{aligned}\bar{K}_\sigma . A = (\sigma(K).S_{p,q}).A &= \sigma(K).(S_{p,q}.A) \\ &= \sigma(K).(qB+(2p+q)A_u) \\ &= q(j_*j^*(\sigma(K))) + (2p+q)(\sigma(K).A_u)\end{aligned}$$

and so $j_*(\bar{K}_\sigma)_{p,q} = j_*j^*(\sigma(K))$. Since j_* is injective, we conclude that $(\bar{K}_\sigma)_{p,q} = j^*(\sigma(K))$.

The strategy for the computation of products will be first to show that $Hal^{\cdot}(W) \otimes \mathbb{Q}$ is generated, as a $\mathbb{Q}$-algebra, by elements of codimension one, and then to show how to compute products of elements of the basis. The following obvious relations will be used throughout:

22.3. $\langle L^i\rangle.\langle L^j\rangle = \langle L^{i+j}\rangle$, $\quad \langle L^i\rangle.\langle \check{L}^j\rangle = \langle L^i\check{L}^j\rangle$, $\quad \langle \check{L}^i\rangle.\langle \check{L}^j\rangle = \langle \check{L}^{i+j}\rangle$.

With a few straightforward computations, from lemma 22.2 we get the following products:

22.4. (1) $\langle S_{p,q}\rangle.\langle L\rangle$ has characters $(2q,4p,0)$ and $2\ell[p,q]$.

(2) $\langle S_{p,q}\rangle.\langle \check{L}\rangle$ has characters $(0,4q,2p)$ and $2\check{\ell}[p,q]$.

(3) $\langle S_{p,q}\rangle.\langle L^2\rangle$ has characters $(4p+2q,4p+2q,8p+8q)^T$ and $4\ell^2[p,q]$.

(4) $\langle S_{p,q}\rangle.\langle \frac{L\check{L}}{2}\rangle$ has characters $(4p+2q,2p+2q,2p+4q)^T$ and $2(\ell^2+\check{\ell}^2)[p,q]$

(5) $\langle S_{p,q}\rangle.\langle \check{L}^2\rangle$ has characters $(8p+8q,2p+4q,2p+4q)^T$ and $4\check{\ell}^2[p,q]$.

(6) $\langle S_{p,q}\rangle.\langle L^3\rangle$ has characters $(4p+2q,8p+4q)^T$ and 0.

(7) $<S_{p,q}>.<\check{L}^3>$ has characters $(4p+8q,2p+4q)^T$ and 0.

(8) $<S_{p,q}>.<L^2\check{L}>$ has characters $(8p+4q,8p+8q)^T$ and $8[p,q]$.

(9) $<S_{p,q}>.<L\check{L}^2>$ has characters $(8p+8q,4p+8q)^T$ and $8[p,q]$.

(10) $<S_{p,q}>.<L^4> = 4p+2q$

(11) $<S_{p,q}>.<L^3\check{L}> = 8p+4q$

(12) $<S_{p,q}>.<L^2\check{L}^2> = 8p+8q$

(13) $<S_{p,q}>.<L\check{L}^3> = 4p+8q$

(14) $<S_{p,q}>.<\check{L}^4> = 2p+4q$ □

To compute the products of $<S_{p,q}>$ with $<S>$, $<T>$, and $<\check{S}>$ (notations of (20.6)) we only need check

$$<S> = <\frac{\check{L}^3}{2}> - <\frac{L\check{L}^2}{4}>$$

$$<\check{S}> = <\frac{L^3}{2}> - <\frac{L^2\check{L}}{4}>$$

$$<T> = <\frac{5}{8}L^2\check{L}> + <\frac{5}{8}\check{L}L^2> - <\frac{3}{4}L^3> - <\frac{3}{4}\check{L}^3>$$

and then compute the products using those equalities of 22.4 that are pertinent. We get:

22.5. (1) $<S_{p,q}>.<S>$ has characters $(2q,0)^T$ and $-2[p,q]$.

(2) $<S_{p,q}>.<\check{S}>$ has characters $(0,2p)^T$ and $-2[p,q]$.

(3) $<S_{p,q}>.<T>$ has characters $(4p,4q)^T$ and $10[p,q]$. □

Again by straightforward computations (using formulae (20.5)-(20.7)) the expression of the above products in terms of the basis turns out to be as follows.

22.6. (1) $<S_{p,q}>.<L> = 2<H_{p,q}> - (4p+2q)<\frac{L\check{L}}{2}>$

(2) $<S_{p,q}>.<\check{L}> = 2<\check{H}_{p,q}> - (2p+4q)<\frac{L\check{L}}{2}>$

(3) $<S_{p,q}>.<L^2> = 4<G_{p,q}>+(4p+2q)<L\check{L}^2-\check{L}^3>$

(4) $<S_{p,q}>.<\check{L}^2> = 4<\check{G}_{p,q}>+(2p+4q)<L^2\check{L}-L^3>$

(5) $<S_{p,q}>.<\frac{L\check{L}}{2}> = 2<G_{p,q}> + 2<\check{G}_{p,q}> - (4p+4q)(<S>+<T>+<\check{S}>)$

(6) $<S_{p,q}>.<L^2\check{L}>= 8<\Gamma_{p,q}> - 4q<\Gamma>$

(7) $<S_{p,q}>.<L\check{L}^2>= 8<\Gamma_{p,q}> - 4p<\check{\Gamma}>$

(8) $<S_{p,q}>.<S> = -2<\Gamma_{p,q}>+ (2p+4q)<\Gamma> + (2p+2q)<\check{\Gamma}>$

(9) $<S_{p,q}>.<\check{S}> = -2<\Gamma_{p,q}>+ (2p+2q)<\Gamma> + (2p+4q)<\check{\Gamma}>$

(10) $<S_{p,q}>.<T> = 10<\Gamma_{p,q}>- (6p+10q)<\Gamma> - (10p+6q)<\check{\Gamma}>$ □

In particular we observe that $Hal^{\bullet}(W) \otimes \mathbb{Q}$ is generated, as $\mathbb{Q}$-algebra, by the basis elements of codimension 1. In fact we have found expressions of the basis elements of codimension 2,3 and 4 as linear combinations with rational coefficients of products of codimension 1 basis elements. Thus our task is to find out the products of any two elements of the codimension 1 basis, and of these only those of the form $<S_{p,q}>.<S_{p',q'}>$ are left. The result of this product is the contents of next theorem.

22.7. Theorem

Assume $p/q \leq p'/q'$. Then the global characteristic numbers of $<S_{p,q}>.<S_{p',q'}>$ are $(2(p+2q)q',\ 2(4p'q-pq'),\ 2p(2p'+q'))$. Moreover, the only pairs of local characteristic number of the cycle $K := S_{p,q} \cap S_{p',q'}$ are (p,q) and (p',q') and the corresponding multiplicity cycles are

$$K_{p,q} = (4p'+2q')\check{\ell}\ , \quad K_{p',q'} = (2p+4q)\ell \quad \text{if} \quad p/q < p'/q'$$

and

$$K_{p,q} = (4p+2q)\check{\ell} + (2p+4q)\ell \quad \text{if} \quad p/q = p'/q' \ .$$

Proof

The global characters of $K := S_{p,q} \cap S_{p',q'}$ can be obtained by first calculating the intersection numbers of K with L^3, $L^2\check{L}$, $L\check{L}^2$, and

$\check{L}^3$. We have that

$$K.L^3 = p(S_{p,q}, S_{p',q'}.L^3) \qquad \text{(by 21.1)}$$
$$= 4(2p+q)(2p'+q') \qquad \text{(formula 20.2)} .$$

Likewise we get

$$K.L^2\check{L} = 8(p+q)(2p'+q') ,$$
$$K.L\check{L}^2 = 8(p+2q)(p'+q') , \quad \text{and}$$
$$K.\check{L}^3 = 4(p+2q)(p'+2q') .$$

Now from these equalities the global characters of K, namely K.S, K.T, and $K.\check{S}$, can be obtained in a straightforward manner, which yields the claimed numbers.

That only (p,q), (p',q') can be local characteristic pairs for K is a consequence of lemma 22.2. Indeed, for any other characteristic pair of K there would exist and excedentary component of $K \cap B$ all of whose conics would go through two general points, or would go through a general point and be tangent to a general line, or would be tangent to two general lines, and so in all cases such a component would have dimension 1, which contradicts it being excedentary.

So next step is the computation of $K_{p,q}$ and $K_{p',q'}$. To do this we first compute **A**.K and $\check{\mathbf{A}}$.K with a method similar to that used at the beginning of this proof. We get

$$\mathbf{A}.K = (16pp'+8pq'+24p'q+12qq', \ 4pq'+8p'q+12qq', \ 8pq'+16qq')^T$$
$$\check{\mathbf{A}}.K = (16pp'+8pq', \ 12pp'+4pq'+8p'q, \ 12pp'+8pq'+24p'q+16qq')^T .$$

At this point we use the fact, which is a direct consequence of the definition of the cycles $K_{p,q}$ (see 18.1), that

$$(*) \qquad \mathbf{A}.K = qj_*K_{p,q} + q'j_*K_{p',q'} + R$$

and

(**) $$\check{A}.K = pj_*K_{p,q} + p'j_*K_{p',q'} + \check{R} ,$$

where R and $\check{R}$ are effective cycles, each a sum of irreducible cycles not contained in **B**.

Actually it turns out that $R = \rho D_Q$ and $R = \check{\rho}\check{D}_v$, where Q is the intersection of the lines u and u' and v is the line joining P and P', where P, P', u, u' are the points and lines used in the definition of $S_{p,q}$ and $S_{p',q'}$ (recall that D_Q is the cycle of pairs of lines with its double point at Q and that $\check{D}_v$ are pairs of points on v), and where ρ and $\check{\rho}$ are positive integers. In fact since $\mathbf{A} \cap S_{p,q} = \mathbf{B} \cup A_u$ (A_u the cycle of double lines with its double point on u), the components appearing in R must be components of $A_u \cap S_{p',q'}$ not contained in **B**. But $A_u \cap S_{p',q'} = B_u \cup D_Q$, so indeed R is of the form ρD_Q, ρ a positive integer. The expression of $\check{R}$ is seen with a similar argument.

Therefore we see that $\mathbf{A}.K - \rho D_Q$ and $\check{\mathbf{A}}.K - \check{\rho}\check{R}$ are in the image of j_*. Since $D_Q \sim (1,0,0)^T$, $\check{D}_v \sim (0,0,1)^T$, and the elements $(a,b,c)^T$ in the image of j_* satisfy $a+c = 2b$, we infer that

$$\mathbf{A}.K - \rho D_Q = (16p'q + 8qq',\ 4pq'+8p'q+16qq',\ 8pq'+16qq')^T$$
$$\check{\mathbf{A}}.K - \check{\rho}\check{D}_v = (16pp'+8pq',\ 12pp'+4pq'+8p'q,\ 8pp'+16p'q)^T .$$

These expressions and the relations (*) and (**) allow us to solve for $K_{p,q}$ and $K_{p',q'}$. If $p/q < p'/q'$ we get, again in a straightforward manner, that

$$j_*K_{p,q} = (4p'+2q')(4,2,0)^T$$

and

$$j_*K_{p',q'} = (2p+4q)(0,2,4)^T ,$$

and hence

$$K_{p,q} = (4p'+2q')\check{\ell}$$
$$K_{p',q'} = (2p+4q)\ell .$$

If $p/q = p'/q'$, then only $K_{p,q}$ is present and it has to be calculated from the relation

$$A.K - \rho D_Q = qj_* K_{p,q} ,$$

which gives that

$$K_{p,q} = (4p+2q)\check{\ell} + (2p+4q)\ell \quad . \quad \square$$

We may also express $<S_{p,q}>.<S_{p',q'}>$ in terms of the basis, using 20.5. We get

22.8. Corollary

If $p/q \leqslant p'/q'$ then

$$<S_{p,q}>.<S_{p',q'}> = -6(p+2q)(2p'+q')<\frac{L\check{L}}{2}> + (4p'+2q')<\check{H}_{p,q}> + (2p+4q)<H_{p',q'}>$$

If we express $\check{H}_{p,q}$ and $H_{p',q'}$ using 22.6, (1) and (2), we get the expression

$$<S_{p,q}>.<S_{p',q'}> = -2(p+2q)(2p'+q')<\frac{L\check{L}}{2}> + (2p'+q')<S_{p,q}>.<\check{L}> +$$
$$+ (p+2q) < S_{p',q'}><L> ,$$

which can be written as

$$(<S_{p,q}>-(p+2q)<L>)(<S_{p',q'}>- (2p'+q')<\check{L}>) = 0 .$$

22.9. Remark

As an easy consequence of the product rules explained in this section, the calculation of local characteristic numbers of a given condition K can be reduced, intersecting K with cycles from among L^{4-i}, $L^{3-i}\check{L},\ldots, \check{L}^{4-i}$ (where i is the codimension of K, that we will assume $\leqslant 3$), to the computation of the local characteristic numbers of a 1-dimensional system of conics, as Halphen already did for the codimension 1 case ([H.3], §13).

If K has codimension 1 and its local characters are $\beta_1[p_1,q_1],\ldots,\beta_s[p_s,q_s]$, then the local characters of $L_P L_Q \check{L}_u K$ (where P, Q, u are chosen generically) are exactly $8\beta_1[p_1,q_1],\ldots,8\beta_s[p_s,q_s]$.

If K has codimension 2 and local characters $(\beta_1^1,\beta_1^2)[p_1,q_1],\ldots,(\beta_s^1,\beta_s^2)[p_s,q_s]$ then $\check{L}_u\check{L}_v K$ has local characters $4\beta_1^1[p_1,q_1],\ldots,4\beta_s^1[p_s,q_s]$ and $L_P L_Q K$ has local characters $4\beta_1^2[p_1,q_1],\ldots,4\beta_s^2[p_s,q_s]$.

Finally, if K has codimension 3 and its local characters are $(\beta_1^1,\beta_1^2)[p_1,q_1]$, ..., $(\beta_s^1,\beta_s^2)[p_s,q_s]$ then $\check{L}_u.K$ has local characters $2\beta_1^1[p_1,q_1],\ldots,2\beta_s^1[p_s,q_s]$, while $L_P.K$ has local characters $2\beta_1^2[p_1,q_1],\ldots$ $\ldots,2\beta_s^2[p_s,q_s]$.

22.10. Remark

Using the table of products explained in this section, one can obtain again Halphen's second formula 14.6 by a rather long computation but not very difficult.

§23. Further examples

Here we give an example of one-dimensional system and an example of third order condition. In both cases we compute their global and local characters. First of all we state a lemma which will be useful in the sequel.

23.1. Lemma

Let $f_i(x) \in \mathbb{C}[[x^{1/p}]]$ be formal power series of orders ℓ_i, $i=1,\ldots,s$, so that $f_i = A_i x^{\ell_i} + \ldots$. Consider the system of homogenous linear equations

$$\sum_1^s a_i f_i(x) = 0$$

$$\sum_1^s a_i f_i'(x) = 0$$

$$\dots \dots \dots \dots$$

$$\sum_1^s a_i f_i^{(s-2)} = 0$$

in the unknowns $a_1,\dots,a_s$. If all the orders ℓ_i are different, then the system has rank s-1 and its solution is given by

$$a_i = \lambda(A_1\dots\hat{A}_i\dots A_s) \prod_{\substack{j<j' \\ j,j'\neq i}} (\ell_{j'}-\ell_j)x^{-\ell_i} + \dots ,$$

where $\lambda\in C((x^{1/p}))$.

Proof

It is a well known result for p=1 (see for instance [E-H], § 1), and the general case follows from this case after a change of variables. □

23.2. Example (Halphen-Zeuthen, [H.1])

Let Δ be an irreducible plane algebraic curve of degree ≥ 3. As it is well known there exists a non-constant rational transformation

$$\begin{aligned} f : \Delta &\longrightarrow W \\ z &\longmapsto C_z \end{aligned}$$

by taking, for a generic $z\in\Delta$, C_z to be the (unique) conic such that $i_z(\Delta.C_z)\geq 5$. We define Γ as the image of Δ under f. This system will be called the system of forth order contact conics to Δ. Next we will compute its local and global characters. For this we need a lemma.

Lemma

The map f is a birrational transformation of Δ to Γ.

Proof

We need only check that f is generically injective and this will be done by showing that z is the unique base point of the pencil of point conics obtained by taking the tangent line to the curve $p(\Gamma)$ at the point $p(C_z)$ in $\mathbb{P}_5$. To show this let $z(t) = (x(t),y(t))$ be an affine parametric expression of a branch δ of Δ with $z(0) = z$.

The conditions for a conic

$$a_{11}x^2 + 2a_{12}xy + a_{22}y^2 + 2a_{1o}x + 2a_{2o}y + a_{oo} = 0$$

to have a four order contact at the point $z(t)$ are

$$F(a_{ij},t) := a_{11}x^2(t)+2a_{12}x(t)y(t)+a_{22}y^2(t)+2a_{1o}x(t)+2a_{2o}y(t)+a_{oo} = 0$$

and

$$\frac{\partial^h F}{\partial t^h}(a_{ij},t) = 0 \ , \qquad 1 \leqslant h \leqslant 4 \ .$$

These equations allow us to determine power series

$$a_{ij} = a_{ij}(t)$$

which give a parametric representation of the branch γ of Γ that corresponds to δ under f. Notice that the line in $\mathbb{P}_5$ spanned by the points $(a_{ij}(0))$ and $(da_{ij}/dt(0))$ is the tangent line to $p(\Gamma)$ at $p(C_z)$.

On the other hand taking derivative with respect to t of the identity

$$F(a_{ij}(t),t) = 0$$

and using the fact that F is linear in the a_{ij} we conclude that

$$F(\frac{da_{ij}}{dt}, t) = 0 \ .$$

Proceeding similarly with the identities

$$\frac{\partial^i F}{\partial t^i}(a_{ij}(t),t) = 0$$

for $i=1,2,3$ we obtain the relations

$$\frac{\partial^i F}{\partial t^i}\left(\frac{da_{ij}}{dt}, t\right) = 0 \quad .$$

These relations, when evaluated at $t=0$, imply that the conic whose coefficients are $da_{ij}/dt\ (0)$ has a third order contact with δ which in turn imply that it has a third order contact with C_z. □

The previous lemma tells us that there is a bijection between branches of Δ and Γ under f. So to study the degenerations of Γ it is enough to take branches δ of Δ and study the local characters of $f(\delta)$. Let

$$y = A\, x^{1+\rho} + \ldots \quad , \qquad \rho > 0, \quad A \neq 0,$$

be the Puiseux expansion of δ in suitable affine coordinates. We will distinguish three cases.

$\rho \neq 1$

In this case lemma 23.1, applied to the system of equations obtained taking the relation

$$a_{11}x^2 + 2a_{12}xy(x) + a_{22}y^2(x) + 2a_{1o}x + 2a_{2o}y(x) + a_{oo} = 0$$

and its first four derivatives with respect to x, yield expressions for the a_{ij}'s as broken power series in x, expressions that themselves allow us to compute the order with respect to x of suitable functions X,Y as defined in §4. The result is that $\mathrm{ord}_x(X) = 2$ and $\mathrm{ord}_x(Y) = 2\rho$. Now if p is the order of δ and q its class, so that $\rho = q/p$, then $f(\delta)$ has order $2q$ and class $2p$.

$\rho = 1$ *and in the Puiseux series of* δ *there is no fractionary exponent less than* 4.

There exist complex numbers B,C such that

$$R = y - Ax^2 - Bxy - Cy^2$$

has order in $x > 4$. Now setting the general equation of a conic in the form

$$(a_{11}+2a_{20}A)x^2 + (2a_{12}+2a_{20}B)xy + (a_{22}+2a_{20}C)y^2 + 2a_{10}x + 2a_{20}R + a_{00} = 0$$

and following a similar procedure as in the previous case one obtains that the center of $f(\delta)$ is the conic $R=0$, which is non-degenerate, so that in this case there are no degenerations.

$\rho=1$ *and there exists a fractionary exponent* $2+\frac{e}{p}$ *in the Puiseux expansion of* δ *such that* $0<\frac{e}{p}<2$, *where* p *is the order of* δ.

There exists a complex number B such that

$$R = y - Ax^2 - Bxy$$

has order in x equal to $2+\frac{e}{p}$. Taking now the equation of the conic as in the previous case, lemma 23.1 allows us to determine a degeneration whose order and class are both equal to $2p-e$.

Once the local characters of Γ are known, the formulae obtained in (5.6) can be used to obtain the global characters of Γ. Summarizing we have:

The system of conics that have a forth order contact with an irreducible plane curve has only the following degenerations:

(a) a degeneration with order $2q$ and class $2p$ for each branch of Δ whose order is p and whose class in q, with $p \neq q$;

(b) a degeneration with order and class both equal to $2p-e$ for each branch of Δ with order and class equal to p and such that the first characteristic exponent in the Puiseux expansion of the branch is of the form $2+\frac{e}{p}$ with $0<\frac{e}{p}<2$.

On the other hand the global characteristic numbers (μ,ν) of Γ are given by the expressions

$$\mu = \frac{2}{3}(P+2Q) + 2P' - E, \qquad \nu = \frac{2}{3}(2P+Q) + 2P' - E,$$

where P and Q are the sums of the integers p and q, respectively, of the branches in (a), and P' and E the sum of the integers p and e of the branches in (b). □

23.3. Example

As in the previous example, let Δ be an irreducible plane algebraic curve of degree $\geqslant 3$. Consider the algebraic correspondence T in $\Delta \times \mathbf{W}$ given by $(z,C)\in T \Leftrightarrow i_z(C,\Delta)\geqslant 4$. Let K denote the strict transform of Δ under the correspondence T. As it is easy to see, K is an (irreducible) condition of order 3. Moreover, for $z\in\Delta$ generic, the transform $T(z)$ is a four-point contact pencil that contains the tangent to Δ at z counted twice with z as a double focus. So $K\cap \mathbf{B}$ is excedentary.

In order to compute the local characteristic numbers of K we apply the method explained in Remark 22.10. Thus we intersect K with L_P and $\check{L}_u$, where P and u are a generic point and a generic line, respectively. Let us consider the case $K.L_P$. If we set $\Gamma = K.L_P$, then an argument similar to that used in the lemma in example 22.3 shows that Γ is birrationally equivalent to Δ. Notice that to a generic $z\in\Delta$ there corresponds the unique conic through P that has a third order contact with Δ at z. So the branches of Γ are in one-to-one correspondence with the branches of Δ and, as in the previous example, we will study the degenerations

γ of Γ in terms of the corresponding branches δ of Δ. We shall distinguish two cases according to whether the tangent to δ goes through P (case (a)) or does not go through P (case (b)).

Case (a)

Since P is generic the order and class of δ are both 1, and so we may choose affine coordinates in which δ has a Puiseux expansion of the form

$$y = Ax^2 + \dots , \qquad A \neq 0,$$

and P has projective coordinates (0,1,0) (point at infinity of the x-axis). The conics we are considering have therefore the form

$$2a_{12}xy + a_{22}y^2 + 2a_{1o}x + 2a_{2o}y + a_{oo} = 0$$

and proceeding as in the previous example one finds a degeneration of Γ whose order and class are both equal to 2. Clearly the number of degenerations is the class of Δ.

Case (b)

In the case choose affine coordinates in such a way that δ has a Puiseux expansion of the form

$$y = Ax^{1+\rho} + \dots , \qquad \rho > 0, \quad A \neq 0,$$

and that P is the point at infinity of the y-axis. In this way the conics through P have the form

$$a_{11}x^2 + 2a_{12}xy + 2a_{1o}x + 2a_{2o}y + a_{oo} = 0 .$$

Now one distinguishes three cases, just as in the previous example, except that in the second and third cases instead of "fractionary exponent less than 4" now we must use "fractionary exponent less than 3". In any of these cases the computations of the local characters are parallel to the corresponding cases in the previous example. Summarizing we have:

The system Γ has only the following degenerations:

(1) m degenerations of order and class both equal to 2, where m is the class of Δ.

(2) a degeneration of order 0 and class $2p+q$ for each branch of Δ of order p and class q, with $p \neq q$.

(3) a degeneration of order 0 and class $3p-3e$ for each branch of Δ whose order and class are both equal to p and whose first characteristic exponent is $2 + \frac{e}{p}$ with $0 < \frac{e}{p} < 1$.

Then the global characteristic numbers (μ, ν) of Γ are given by the expressions

$$\mu = 2m + \frac{2P+Q}{3} + P'-E \quad , \qquad \nu = 2m + \frac{4P+2Q}{3} + 2P' - 2E \quad ,$$

where P and Q are the sums of the integers p and q appearing in (2) and P' and E are the sums of the integers p and e appearing in (3).

The characters of the intersection $K.\check{L}_u$ are determined by duality and are as follows:

(1) n degenerations of order and class both equal to 2, where n is the order of Δ.

(2) a degeneration of order $p+2q$ and class 0 for each branch of Δ of order p and class q with $p \neq q$.

(3) a degeneration of order $3p-3e$ and class 0 for each branch of Δ of order and class both equal to p and whose first characteristic exponent is $2 + \frac{e}{p}$ with $0 < \frac{e}{p} < 1$.

Then the global characteristic numbers (μ',ν') of $K.\check{L}_u$ are given by the expressions

$$\nu' = 2n + \frac{P+2Q}{3} + P' - E\,, \qquad \mu' = 2n + \frac{2P+4Q}{3} + 2P' - 2E\,.$$

Applying remark 22.10 to K we can now easily determine the local characters of K, which turn out to be a single characteristic pair (1,1) with multiplicity (m,n).

On the other hand the global characteristic numbers of K are

$$K.L^2 = \mu\,, \qquad K.(\tfrac{1}{2}\, L\check{L}) = \nu = \mu'\,, \qquad K.\check{L}^2 = \nu'.$$

References.

B **Beauville, A.**
Variétés de Prym et Jacobiennes Intermédiaires.
Ann. Sci. E.N.S. 10 (1977) pp. 304-379.

Ca.1 **Casas, E.**
Singularidades unidimensionales de una superficie algebraica.
Coll. Math. XXIX, 1 (1978), pp.21-42.

Ca.2 Singularidades de una hoja de superficie algebraica a partir de su serie de Puiseux.
Coll. Math. XXIX, 2 (1978) pp. 133-165.

C.1 **Chasles, M.**
Détermination du nombre de sections coniques qui doivent toucher cinq courbes données d'ordre quelconque, ou satisfaire à diverses autres conditions.
C.R.A.Sc. 58 (1864) pp. 297-308.

C.2 Construction des coniques qui satisfont à cinq conditions. Nombre des solutions dans chaque question.
C.R.A.Sc. 58 (1864) pp. 297-308.

C.3 Systèmes de coniques qui coupent des coniques données sous des angles donnés, ou sous des angles indéterminés mais dont les bisectrices ont les directions données.
C.R.A.Sc. 58 (1864) pp. 425-431.

C.4 Considérations sur la méthode générale exposée dans la séance du 15 février. Différences entre cette méthode et la méthode analytique. Procédés généraux de demonstration.
C.R.A.Sc. 58 (1864) pp. 1167-1175.

C.5 Exemples des procédés de demonstration annoncés dans la séance précédente.
C.R.A.Sc. 59 (1864) pp. 7-15.

C.6 Suite des propriétés relatives aux systèmes de sections coniques.
C.R.A.Sc. 59 (1864) pp. 93-97.

C.7. Questions dans lesquelles il y a lieu de tenir compte des points singuliers des courbes d'ordre supérieur. Formules générales comprenant la solution de toutes les questions relatives aux sections coniques.
C.R.A.Sc. 59 (1864) pp. 209-218.

C.8 Questions dans lesquelles entrent des conditions multiples, telles que des conditions de double contact ou de contact d'ordre supérieur.
C.R.A.Sc. 59 (1864) pp. 346-357.

C.9 Observations relatives à la théorie des systèmes de courbes.
C.R.A.Sc. 63 (1866) pp. 816-821.

C.10 Propriétés des systèmes de coniques relatives, toutes, à certaines séries de normales en rapport avec d'autres lignes ou divers points.
C.R.A.Sc. 72 (1871) pp. 419-430.

C.11 Propriétés des systèmes de coniques, dans lesquels se trouvent des conditions de perpendicularité entre diverses séries de droites.
C.R.A.Sc. 72 (1871) pp. 487-494.

C.12 Théorèmes divers concernant les systèmes de coniques représentés par deux caractéristiques.
C.R.A.Sc. 72 (1871) pp. 511-520.

C.13 Propriétés des courbes d'ordre et classe quelconques démontrés par le Principe de correspondence.
C.R.A.Sc. 72 (1871) pp. 577-588.

DC-P **DeConcini, C.; Procesi, C.**
Complete Symmetric Varieties II (Intersection theory).
Preprint (1982), Istituto "Guido Castelnuovo", Roma.

E-H **Eisenbud, D. - Harris, J.**
Divisors on general curves and cuspidal rational curves.
Invent. math. 74, (1983) pp. 371-418.

F **Fulton, W.**
Intersection Theory.
Springer-Verlag (1984), Ergebnisse 3. Folge, Band 2.

H.1 **Halphen, G.H.**
Sur les caractéristiques des systèmes de coniques et des surfaces de second ordre.
Journal de l'École Polytechnique de Paris, 45e cahier XXVIII (1878) pp. 27-84. Oeuvre II. Gauthier-Villars, Paris (1918) pp. 1-57.

H.2 Théorie des caractéristiques pour les coniques.
Proc. London Math. Soc. IX (1877-78) pp. 149-177. Oeuvre II Gauthier-Villar, Paris (1918), pp. 58-92.

H.3 Sur le nombre des coniques qui dans un plan satisfont à cinq conditions projectives et indépendentes entre elles.
Proc. London Math. Soc. X (1878-79) pp. 76-91. Oeuvres II, Gauthier-Villars, Paris (1918), 275-289.

H.4. Sur les caractéristiques dans la théorie des coniques sur le plan et dans l'espace et des surfaces du second ordre.
C.R.A.Sc. 76 (1873) pp. 1074-1077. Oeuvres I, Gauthier-Villars, Paris (1916) pp. 159-162.

Har **Hartshorne, R.**
Algebraic Geometry.
Springer-Verlag, New-York (1977).

J.1 **de Jonquières, J.P.E.F.**
Théorèmes généraux concernant les courbes géométriques planes d'un ordre quelconque.
J. Math. Pures et Appl. (2) 6 (1861) pp. 113-134.

J.2 Sur la détermination des valeurs des caractéristiques dans les séries ou systèmes élémentaires de courbes et de surfaces.
C.R.A.Sc. 63 (1966) pp. 793-797.

J.3 Mémoire sur les contacts multiples d'ordre quelconque des courbes de degrée r qui satisfont à des conditions données, avec une courbe fixe du degré m; suivi de quelques réflexions sur la solution d'un grand nombre de questions concernant les propriétés des courbes et des surfaces algébriques.
Journ. Crelle 66 (1866) pp. 289-321.

K.1 **Kleiman, S.L.**
The transversality of a general translate.
Compos. Math. 28 (1974) pp. 287-297.

K.2 Chasles's enumerative theory of conics. A historical introduction.
in Studies in algebraic geometry. Studies in Math. 20, M.A.A. (1980).

N **Northcott, D.G.**
A general theory of one-dimensional local rings.
Proc. Glasgow Math. Ass. 2 (1956).

S **Shafarevich, I.R.**
Basic algebraic geometry.
Grundlehren 213, Springer-Verlag, Heidelberg (1974).

S-K **Semple, J.G - Kneebone, G.T.**
Algebraic projective geometry.
Oxford Univ. Press, London (1952).

S-S **Shreirer, O. - Sperner, E.**
Projective Geometry.
Chelsea Pub. Co. New-York (1961).

W **van der Waerden, B.L.**
Zur algebraischen Geometrie XV. Lösung des Charakteristikenproblems für Kegelschnitte.
Math. Ann. 115 (1938) pp. 645-655.

Wal **Walker, R.J.**
Algebraic curves.
Princeton Univ. Press, 1950.

Index of terms

Index of symbols

Vol. 1090: Differential Geometry of Submanifolds. Proceedings, 1984. Edited by K. Kenmotsu. VI, 132 pages. 1984.

Vol. 1091: Multifunctions and Integrands. Proceedings, 1983. Edited by G. Salinetti. V, 234 pages. 1984.

Vol. 1092: Complete Intersections. Seminar, 1983. Edited by S. Greco and R. Strano. VII, 299 pages. 1984.

Vol. 1093: A. Prestel, Lectures on Formally Real Fields. XI, 125 pages. 1984.

Vol. 1094: Analyse Complexe. Proceedings, 1983. Edité par E. Amar, R. Gay et Nguyen Thanh Van. IX, 184 pages. 1984.

Vol. 1095: Stochastic Analysis and Applications. Proceedings, 1983. Edited by A. Truman and D. Williams. V, 199 pages. 1984.

Vol. 1096: Théorie du Potentiel. Proceedings, 1983. Edité par G. Mokobodzki et D. Pinchon. IX, 601 pages. 1984.

Vol. 1097: R.M. Dudley, H. Kunita, F. Ledrappier, École d'Éte de Probabilités de Saint-Flour XII – 1982. Edité par P.L. Hennequin. X, 396 pages. 1984.

Vol. 1098: Groups – Korea 1983. Proceedings. Edited by A.C. Kim and B.H. Neumann. VII, 183 pages. 1984.

Vol. 1099: C.M. Ringel, Tame Algebras and Integral Quadratic Forms. XIII, 376 pages. 1984.

Vol. 1100: V. Ivrii, Precise Spectral Asymptotics for Elliptic Operators Acting in Fiberings over Manifolds with Boundary. V, 237 pages. 1984.

Vol. 1101: V. Cossart, J. Giraud, U. Orbanz, Resolution of Surface Singularities. Seminar. VII, 132 pages. 1984.

Vol. 1102: A. Verona, Stratified Mappings – Structure and Triangulability. IX, 160 pages. 1984.

Vol. 1103: Models and Sets. Proceedings, Logic Colloquium, 1983, Part I. Edited by G.H. Müller and M.M. Richter. VIII, 484 pages. 1984.

Vol. 1104: Computation and Proof Theory. Proceedings, Logic Colloquium, 1983, Part II. Edited by M.M. Richter, E. Börger, W. Oberschelp, B. Schinzel and W. Thomas. VIII, 475 pages. 1984.

Vol. 1105: Rational Approximation and Interpolation. Proceedings, 1983. Edited by P.R. Graves-Morris, E.B. Saff and R.S. Varga. XII, 528 pages. 1984.

Vol. 1106: C.T. Chong, Techniques of Admissible Recursion Theory. IX, 214 pages. 1984.

Vol. 1107: Nonlinear Analysis and Optimization. Proceedings, 1982. Edited by C. Vinti. V, 224 pages. 1984.

Vol. 1108: Global Analysis – Studies and Applications I. Edited by Yu.G. Borisovich and Yu.E. Gliklikh. V, 301 pages. 1984.

Vol. 1109: Stochastic Aspects of Classical and Quantum Systems. Proceedings, 1983. Edited by S. Albeverio, P. Combe and M. Sirugue-Collin. IX, 227 pages. 1985.

Vol. 1110: R. Jajte, Strong Limit Theorems in Non-Commutative Probability. VI, 152 pages. 1985.

Vol. 1111: Arbeitstagung Bonn 1984. Proceedings. Edited by F. Hirzebruch, J. Schwermer and S. Suter. V, 481 pages. 1985.

Vol. 1112: Products of Conjugacy Classes in Groups. Edited by Z. Arad and M. Herzog. V, 244 pages. 1985.

Vol. 1113: P. Antosik, C. Swartz, Matrix Methods in Analysis. IV, 114 pages. 1985.

Vol. 1114: Zahlentheoretische Analysis. Seminar. Herausgegeben von E. Hlawka. V, 157 Seiten. 1985.

Vol. 1115: J. Moulin Ollagnier, Ergodic Theory and Statistical Mechanics. VI, 147 pages. 1985.

Vol. 1116: S. Stolz, Hochzusammenhängende Mannigfaltigkeiten und ihre Ränder. XXIII, 134 Seiten. 1985.

Vol. 1117: D.J. Aldous, J.A. Ibragimov, J. Jacod, Ecole d'Ét Probabilités de Saint-Flour XIII – 1983. Édité par P.L. Henne IX, 409 pages. 1985.

Vol. 1118: Grossissements de filtrations: exemples et applica Seminaire, 1982/83. Edité par Th. Jeulin et M. Yor. V, 315 pages. 1

Vol. 1119: Recent Mathematical Methods in Dynamic Programn Proceedings, 1984. Edited by I. Capuzzo Dolcetta, W.H. Fle and T. Zolezzi. VI, 202 pages. 1985.

Vol. 1120: K. Jarosz, Perturbations of Banach Algebras. V, 118 p 1985.

Vol. 1121: Singularities and Constructive Methods for Their Treatr Proceedings, 1983. Edited by P. Grisvard, W. Wendland and Whiteman. IX, 346 pages. 1985.

Vol. 1122: Number Theory. Proceedings, 1984. Edited by K. Al VII, 217 pages. 1985.

Vol. 1123: Séminaire de Probabilités XIX 1983/84. Proceedings. E par J. Azéma et M. Yor. IV, 504 pages. 1985.

Vol. 1124: Algebraic Geometry, Sitges (Barcelona) 1983. Proc ings. Edited by E. Casas-Alvero, G.E. Welters and S. Xa Descamps. XI, 416 pages. 1985.

Vol. 1125: Dynamical Systems and Bifurcations. Proceedings, 1 Edited by B.L.J. Braaksma, H.W. Broer and F. Takens. V, 129 p 1985.

Vol. 1126: Algebraic and Geometric Topology. Proceedings, 1 Edited by A. Ranicki, N. Levitt and F. Quinn. V, 523 pages. 198

Vol. 1127: Numerical Methods in Fluid Dynamics. Seminar. Edite F. Brezzi, VII, 333 pages. 1985.

Vol. 1128: J. Elschner, Singular Ordinary Differential Operators Pseudodifferential Equations. 200 pages. 1985.

Vol. 1129: Numerical Analysis, Lancaster 1984. Proceedings. E by P.R. Turner. XIV, 179 pages. 1985.

Vol. 1130: Methods in Mathematical Logic. Proceedings, 1983. E by C.A. Di Prisco. VII, 407 pages. 1985.

Vol. 1131: K. Sundaresan, S. Swaminathan, Geometry and Nonli Analysis in Banach Spaces. III, 116 pages. 1985.

Vol. 1132: Operator Algebras and their Connections with Topo and Ergodic Theory. Proceedings, 1983. Edited by H. Araki, (Moore, Ş. Strătilă and C. Voiculescu. VI, 594 pages. 1985.

Vol. 1133: K.C. Kiwiel, Methods of Descent for Nondifferentiable C mization. VI, 362 pages. 1985.

Vol. 1134: G.P. Galdi, S. Rionero, Weighted Energy Methods in F Dynamics and Elasticity. VII, 126 pages. 1985.

Vol. 1135: Number Theory, New York 1983–84. Seminar. Editec D.V. Chudnovsky, G.V. Chudnovsky, H. Cohn and M.B. Nathans V, 283 pages. 1985.

Vol. 1136: Quantum Probability and Applications II. Proceedir 1984. Edited by L. Accardi and W. von Waldenfels. VI, 534 pag 1985.

Vol. 1137: Xiao G., Surfaces fibrées en courbes de genre deux. IX, 1 pages. 1985.

Vol. 1138: A. Ocneanu, Actions of Discrete Amenable Groups on Neumann Algebras. V, 115 pages. 1985.

Vol. 1139: Differential Geometric Methods in Mathematical Phys Proceedings, 1983. Edited by H. D. Doebner and J. D. Hennig. VI, 3 pages. 1985.

Vol. 1140: S. Donkin, Rational Representations of Algebraic Grou VII, 254 pages. 1985.

Vol. 1141: Recursion Theory Week. Proceedings, 1984. Edited H.-D. Ebbinghaus, G.H. Müller and G.E. Sacks. IX, 418 pages. 198

Vol. 1142: Orders and their Applications. Proceedings, 1984. Edi by I. Reiner and K. W. Roggenkamp. X, 306 pages. 1985.

Vol. 1143: A. Krieg, Modular Forms on Half-Spaces of Quaternio XIII, 203 pages. 1985.

Vol. 1144: Knot Theory and Manifolds. Proceedings, 1983. Editec D. Rolfsen. V, 163 pages. 1985.

1034: J. Musielak, Orlicz Spaces and Modular Spaces. V, 2 pages. 1983.

1035: The Mathematics and Physics of Disordered Media. eedings, 1983. Edited by B.D. Hughes and B.W. Ninham. 432 pages. 1983.

1036: Combinatorial Mathematics X. Proceedings, 1982. Edited R.A. Casse. XI, 419 pages. 1983.

1037: Non-linear Partial Differential Operators and Quantization edures. Proceedings, 1981. Edited by S.I. Andersson and . Doebner. VII, 334 pages. 1983.

. 1038: F. Borceux, G. Van den Bossche, Algebra in a Localic os with Applications to Ring Theory. IX, 240 pages. 1983.

. 1039: Analytic Functions, Błażejewko 1982. Proceedings. Edited J. Ławrynowicz. X, 494 pages. 1983

. 1040: A. Good, Local Analysis of Selberg's Trace Formula. III, 3 pages. 1983.

. 1041: Lie Group Representations II. Proceedings 1982–1983. ted by R. Herb, S. Kudla, R. Lipsman and J. Rosenberg. IX, 0 pages. 1984.

. 1042: A. Gut, K. D. Schmidt, Amarts and Set Function Processes. 258 pages. 1983.

. 1043: Linear and Complex Analysis Problem Book. Edited by Havin, S.V. Hruščëv and N.K. Nikol'skii. XVIII, 721 pages. 1984.

1044: E. Gekeler, Discretization Methods for Stable Initial Value oblems. VIII, 201 pages. 1984.

. 1045: Differential Geometry. Proceedings, 1982. Edited by M. Naveira. VIII, 194 pages. 1984.

. 1046: Algebraic K-Theory, Number Theory, Geometry and Analy-. Proceedings, 1982. Edited by A. Bak. IX, 464 pages. 1984.

l. 1047: Fluid Dynamics. Seminar, 1982. Edited by H. Beirão da iga. VII, 193 pages. 1984.

. 1048: Kinetic Theories and the Boltzmann Equation. Seminar, 81. Edited by C. Cercignani. VII, 248 pages. 1984.

l. 1049: B. Iochum, Cônes autopolaires et algèbres de Jordan. , 247 pages. 1984.

l. 1050: A. Prestel, P. Roquette, Formally p-adic Fields. V, 167 pages. 84.

l. 1051: Algebraic Topology, Aarhus 1982. Proceedings. Edited by Madsen and B. Oliver. X, 665 pages. 1984.

l. 1052: Number Theory, New York 1982. Seminar. Edited by D.V. hudnovsky, G.V. Chudnovsky, H. Cohn and M.B. Nathanson. V, 309 ages. 1984.

l. 1053: P. Hilton, Nilpotente Gruppen und nilpotente Räume. 221 pages. 1984.

l. 1054: V. Thomée, Galerkin Finite Element Methods for Parabolic roblems. VII, 237 pages. 1984.

l. 1055: Quantum Probability and Applications to the Quantum heory of Irreversible Processes. Proceedings, 1982. Edited by Accardi, A. Frigerio and V. Gorini. VI, 411 pages. 1984.

ol. 1056: Algebraic Geometry. Bucharest 1982. Proceedings, 1982. dited by L. Bădescu and D. Popescu. VII, 380 pages. 1984.

l. 1057: Bifurcation Theory and Applications. Seminar, 1983. Edited y L. Salvadori. VII, 233 pages. 1984.

l. 1058: B. Aulbach, Continuous and Discrete Dynamics near anifolds of Equilibria. IX, 142 pages. 1984.

ol. 1059: Séminaire de Probabilités XVIII, 1982/83. Proceedings. dité par J. Azéma et M. Yor. IV, 518 pages. 1984.

ol. 1060: Topology. Proceedings, 1982. Edited by L. D. Faddeev and .A. Mal'cev. VI, 389 pages. 1984.

l. 1061: Séminaire de Théorie du Potentiel. Paris, No. 7. Pro-eedings. Directeurs: M. Brelot, G. Choquet et J. Deny. Rédacteurs: Hirsch et G. Mokobodzki. IV, 281 pages. 1984.

Vol. 1062: J. Jost, Harmonic Maps Between Surfaces. X, 133 pages. 1984.

Vol. 1063: Orienting Polymers. Proceedings, 1983. Edited by J.L. Ericksen. VII, 166 pages. 1984.

Vol. 1064: Probability Measures on Groups VII. Proceedings, 1983. Edited by H. Heyer. X, 588 pages. 1984.

Vol. 1065: A. Cuyt, Padé Approximants for Operators: Theory and Applications. IX, 138 pages. 1984.

Vol. 1066: Numerical Analysis. Proceedings, 1983. Edited by D.F. Griffiths. XI, 275 pages. 1984.

Vol. 1067: Yasuo Okuyama, Absolute Summability of Fourier Series and Orthogonal Series. VI, 118 pages. 1984.

Vol. 1068: Number Theory, Noordwijkerhout 1983. Proceedings. Edited by H. Jager. V, 296 pages. 1984.

Vol. 1069: M. Kreck, Bordism of Diffeomorphisms and Related Topics. III, 144 pages. 1984.

Vol. 1070: Interpolation Spaces and Allied Topics in Analysis. Proceedings, 1983. Edited by M. Cwikel and J. Peetre. III, 239 pages. 1984.

Vol. 1071: Padé Approximation and its Applications, Bad Honnef 1983. Prodeedings. Edited by H. Werner and H.J. Bünger. VI, 264 pages. 1984.

Vol. 1072: F. Rothe, Global Solutions of Reaction-Diffusion Systems. V, 216 pages. 1984.

Vol. 1073: Graph Theory, Singapore 1983. Proceedings. Edited by K.M. Koh and H.P. Yap. XIII, 335 pages. 1984.

Vol. 1074: E.W. Stredulinsky, Weighted Inequalities and Degenerate Elliptic Partial Differential Equations. III, 143 pages. 1984.

Vol. 1075: H. Majima, Asymptotic Analysis for Integrable Connections with Irregular Singular Points. IX, 159 pages. 1984.

Vol. 1076: Infinite-Dimensional Systems. Proceedings, 1983. Edited by F. Kappel and W. Schappacher. VII, 278 pages. 1984.

Vol. 1077: Lie Group Representations III. Proceedings, 1982–1983. Edited by R. Herb, R. Johnson, R. Lipsman, J. Rosenberg. XI, 454 pages. 1984.

Vol. 1078: A.J.E.M. Janssen, P. van der Steen, Integration Theory. V, 224 pages. 1984.

Vol. 1079: W. Ruppert. Compact Semitopological Semigroups: An Intrinsic Theory. V, 260 pages. 1984

Vol. 1080: Probability Theory on Vector Spaces III. Proceedings, 1983. Edited by D. Szynal and A. Weron. V, 373 pages. 1984.

Vol. 1081: D. Benson, Modular Representation Theory: New Trends and Methods. XI, 231 pages. 1984.

Vol. 1082: C.-G. Schmidt, Arithmetik Abelscher Varietäten mit komplexer Multiplikation. X, 96 Seiten. 1984.

Vol. 1083: D. Bump, Automorphic Forms on GL (3,IR). XI, 184 pages. 1984.

Vol. 1084: D. Kletzing, Structure and Representations of Q-Groups. VI, 290 pages. 1984.

Vol. 1085: G.K. Immink, Asymptotics of Analytic Difference Equations. V, 134 pages. 1984.

Vol. 1086: Sensitivity of Functionals with Applications to Engineering Sciences. Proceedings, 1983. Edited by V. Komkov. V, 130 pages. 1984

Vol. 1087: W. Narkiewicz, Uniform Distribution of Sequences of Integers in Residue Classes. VIII, 125 pages. 1984.

Vol. 1088: A.V. Kakosyan, L.B. Klebanov, J.A. Melamed, Characterization of Distributions by the Method of Intensively Monotone Operators. X, 175 pages. 1984.

Vol. 1089: Measure Theory, Oberwolfach 1983. Proceedings. Edited by D. Kölzow and D. Maharam-Stone. XIII, 327 pages. 1984.